CONTENTS

LIST OF TABLES

ACRONYMS AND ABBREVIATIONS

AIDS	almost ideal demand system
CAP	Common Agricultural Policy
CS	consumer surplus
CSERGE	Centre for Social and Economic Research on the Global Climate
CSO	Central Statistical Office
DOE	Department of Employment
ESRC	Economic and Social Research Council
GCSE	General Certificate of Secondary Education
GDP	gross domestic product
GNP	gross national product
GP	general practitioner
HPF	household production function
ICP	International Comparisons Project
IEQ	income evaluation question
IIA	independence of irrelevant alternatives
IPCC	Intergovernmental Panel on Climate Change
IPS	international passenger survey
LES	linear expenditure system
PPP	purchasing power parity
PTCM	pooled travel cost model
QES	quadratic expenditure system
RUM	random utility model
SMSA	Standard Metropolitan Statistical Area
UNDP	United Nations Development Programme

Acknowledgements

The author would like to thank his supervisors at Strathclyde University, Anthony Clunies-Ross and Roger Perman, as well as his de facto supervisor in University College London, David Pearce. Many other individuals have contributed to the completion of this book on an academic level. These include: Richard Blundell, Ian Carson, Graham Herbert, Mike Hulme, Phil Jones, Mick Kelly, Charles Kolstadt, Tim Lloyd, Rob Mendelsohn, Bill Nordhaus, Tim O'Riordan, Simon Ogden, Ana Oliveira, David Parker, Martin Parry, Peter Pearson, Kim Swales, Jim Symons, Richard Tol, Kerry Turner, Alistair Ulph, David Ulph and David Viner. Thanks also to Alex, Anna, Annamaria, Angie, Brett, Ed, Helena, Jane, Janet, Joanne, Kirk, Laura, Mikki, Robert, Salma, Sam and Simona. The financial support of the Economic and Social Research Council (ESRC) is gratefully acknowledged.

The author expresses his particular thanks to Andrea Bigano, Bernard van Praag, Paul Frijters and Robert Mendelsohn for their contributions to this book.

FOREWORD

At the dawn of the 21st century, the impact of human beings on the environment continues to provide cause for concern. For some, what matters is the loss of so much that is beautiful and valuable in its own right. For others, the concern is with the effects of environmental degradation on human health and well-being. Whatever one's view, humans, like all living species, are remarkably adaptive. While this message should be familiar, all too many of the participants in the environmental debate seem unaware of it.

David Maddison demonstrates brilliantly the existence of adaptive responses. He chooses climate change and uses econometric techniques to test a number of fundamental hypotheses. Then, pursuing some leads from the early literature, he shows that people do respond to climate. Simply put, we humans like warmer weather and dislike cold weather, but not without limits. There is some 'optimum' temperature, or set of temperatures, that meets our demand for amenity. Maddison shows this both in terms of the demand for commodities and the demand for tourism. We change our demand for goods and services as climate changes. Indeed, we can use this observation to infer the value that individuals place on climate change.

Does this mean that global warming is not a problem after all? That people will simply adapt to whatever happens? Far from it, but the work of Maddison does point towards some serious weaknesses in the kinds of models used so far to assess the costs and benefits of climate change. First, there are benefits as well as costs from global warming. Second, that profile of costs and benefits changes with temperature. Third, humans adapt, and unless adaptation is built into the integrated assessment models of climate change we risk seriously mis-stating the problem. Maddison argues that the direction of bias will be one of overstating the costs of global warming. Such a demonstration does not alter the formidable distributional problems in climate change: already the vulnerable may be more at risk simply because the ability to adapt reflects existing resource endowments.

The work of David Maddison and that of the other contributors to this book is an example of how statistical technique, managed in a sophisticated way and with an imaginative approach, can illuminate some of the questions that need to be asked about climate change and environmental degradation in particular. Our institution is proud to have been the main spawning-ground for David's work.

Professor David Pearce
Centre for Social and Economic Research on the Global Environment
Department of Economics
University College London

NOTES ON THE AUTHOR

David Maddison is a Senior Research Fellow in the Centre for Social and Economic Research on the Global Environment and Associate Lecturer in the Department of Economics, University College London. He holds two master's degrees and a PhD in Economics from the University of Strathclyde. In a former incarnation he was a research assistant in the Industry Department for Scotland. Apart from the amenity value of climate and economic impacts of climate change, he is also interested in quantifying and valuing the health impacts of air pollution and in managing historical landscapes threatened by infrastructure developments.

INTRODUCTION

The motivation for the research on which this book is based was dissatisfaction with the way in which potential damages from future climate change were portrayed in the literature. Damage estimates failed to use existing valuation methodologies, did not take account of the existing situation in different countries and above all did not account for the fundamental issue of human adaptation to changes in climate.

The main theme of this book is that climate is an important input to many household activities. It determines space heating and cooling requirements. It also affects clothing and nutritional requirements. Climate delimits recreational activities, and particular types of climate promote health and a sense of psychological well-being. Households reveal their implicit valuation of particular types of climate in diverse ways: where they locate themselves, their patterns of consumption, their own assessment of the cost of living. In the case of tourists, preferences for particular types of climates can be inferred from analysis of the travel costs incurred, and in the case of farmers the productivity of particular types of climate is contained in agricultural land prices.

The ability to put a price on the value of small changes in climate is important. The problem is that society is confronted with the need to balance the costs of abating those activities that contribute to climate change against the benefits in terms of damage avoided. And every dollar spent on abatement of greenhouse gas is a dollar not spent on present day development such as education, sanitation and clean water. Already governments are switching funds formerly intended for development assistance into projects intended either to abate or to sequester carbon and one wonders whether welfare is improved as a consequence. We have shadow prices for carbon to help us determine those carbon abatement projects that represent value for money and those which do not. But these shadow prices are derived on the basis of evidence that is unconvincing and in need of revision. We must allocate resources in a way that gives due account to the ability of future households and other sectors to adapt to climate change. If we exaggerate the impacts of climate change or wantonly underestimate the ability of future generations to adapt then we are cheating ourselves.

The basic approach employed by these valuation methodologies is to look for current day analogues to future climates and to measure differences in welfare levels once long run cost-minimizing adaptation has occurred. This begs the question of whether autonomous adaptation may be expected to occur or whether governments should intervene. Households and other sectors will adapt in the end because they have a financial incentive to do so. It is likely that they will adopt the lifestyle and customs of households already residing in particular climatic regions (maybe even in the same country). It is, moreover, difficult to see how governments can hasten the process of adaptation (unless such adaptation requires altering the supply of public goods such as sea defences). The only alternative to the use of climatic analogues is to try to model all possible adaptations that might eventually be made in situ – a mammoth task.

While accepting that adaptation to climate change is possible, one may nevertheless disagree that complete adaptation will occur within the time available. We are not starting from scratch and durable goods and capital goods are often designed to operate in a particular climate. In such circumstances, there may be significant costs depending upon their inability to function as intended in a changed climate. Society would in effect be maladapted to living in that climate. But the basic point here is that although climate change is occurring at a rate that is unprecedented in geological terms, it is still slow in comparison to the time scales relevant to households and to economic activity. An entire wardrobe of clothes might be replaced within a couple of years. Household white goods have a lifetime of perhaps ten years. Other goods like the housing stock, office buildings and infrastructure might take longer but their useful lifetime is still short in comparison to the pace of climate change. Of course while economic sectors can adapt to climate change, natural habitats whose ability to migrate is limited might not. But in the very long run even preferences might change and whereas current generations may like what they know, future generations may not miss what they have never had.

In a sense, we also choose our exposure to extreme events. Structures are built with a mind to withstanding weather anomalies of certain strength. These are calculated by return periods that summarize knowledge based on historical records of the frequency that an event exceeding the critical design dimensions of a structure will occur. If these events become more frequent then society will adapt to them by building higher sea walls or retreating from low-lying areas or constructing stronger buildings in more sheltered locations; and it may even decide to accept a higher degree of risk than before. All this is not to say that changes in the frequency of extreme events do not have a cost attached to them, but rather that it is likely to be a cost which is significantly lower than the option of not responding. And the frequency of some extreme events might actually decrease.

The results of the empirical analyses contained in this book indicate that climate change does not impose uniform losses on all countries. More specifically, when long-run adaptation is accounted for, some countries may stand to benefit considerably from climate change whilst other countries lose. Even this will no doubt be seen as a heretical view by some who regard climate change as an unalloyed bad. Certainly the author has not attempted to perform any kind of global cost benefit analysis of halting climate change and this book considers only a subset of the impacts of climate change. The author nevertheless believes that changes in amenity values are likely to be a large element of any final calculus and that the main challenge confronting policy makers is to commit themselves to assisting those countries that are negatively affected by climate change.

In Chapter 1 a theoretical model is developed to illustrate how household amenity values can become capitalized into both land prices and wage rates. This model is based on the hedonic approach that assumes that migration eliminates the net advantages of different locations through adjustments in land prices and wage rates. Using a data set relating to 127 English counties, Scottish and Welsh regions, Metropolitan areas and London boroughs, the chapter tests the ability of a suite of amenity variables (including climate variables) to explain regional variations in property prices and wage rates. It is shown that British households regard higher temperatures as an amenity and higher precipitation as a disamenity.

A basic assumption of the hedonic technique is that there are no barriers to mobility that prevent prices changing to reflect the net benefits of a given location. But climate variables are undeviating over relatively large distances and the absence of a common language and/or cultural ties may prevent the net advantages associated with a particular region from being eliminated. Methods alternative to the hedonic approach may be required to estimate the amenity value of climate. Accordingly, Chapter 2 seeks to undertake a systematic examination of the role played by climate in determining consumption patterns using cross-sectional data from 60 different countries and then proceeds to calculate the implicit price of a range of climate variables for each of them. It is demonstrated that the amenity value of climate change depends upon the current climate a particular country happens to have.

Chapter 3 considers the impact of climate change on agricultural productivity in Britain, once more using the hedonic approach. In this approach sale price differentials between land characterized by different climates are given an interpretation in terms of underlying productivity differences. Such an approach differs radically from the more conventional 'land allocation' approach to determining the impact of climate change on agriculture. Data characterizing over 400 separate transactions in farmland are analysed and the value of marginal changes in climate variables computed. The study suggests that the financial value of climate variables to farmers could, in some cases, be quite high and also that seasonal patterns are very important. Thus the impact of climate change on British agriculture is likely to depend acutely on differential changes across the seasons. Adjusting these implicit prices to account for the price support given to agricultural commodities would presumably reduce the perceived impact of climate change on agriculture. The opportunity also arises to test a number of auxiliary hypotheses not connected with climate change. These hypotheses concern the ability of landowners to repackage their land, the impact of regulated tenancies on farm prices and the accuracy of the valuations performed by land agents.

Hedonic models only capture the amenity value of climate as experienced at the place of home and work. But what about the amenity value of climate when individuals are travelling, possibly abroad? Chapter 4 investigates the impact of climate change on the chosen destinations of British tourists. Destinations are characterized in terms of climate variables, travel costs and accommodation costs. These and other variables are used to explain the observed pattern of overseas travel in terms of a model based on the precepts of utility maximization. This approach permits the tradeoffs between climate and expenditure to be analysed and effectively identifies the optimal climate for generating tourism. The impact of particular climate change scenarios on typical tourist destinations is predicted.

Chapter 5 is co-written with Andrea Bigano (now with the Catholic University of Leuven) and uses the hedonic modelling approach in the context of Italy. Estimates of the marginal willingness to pay for small changes in the levels of climate amenities are derived using the hedonic technique applied to five consecutive years of house price data in Italy. Hedonic price models are specified in terms of both annual as well as in terms of January and July averages. There is shown to be considerable empirical support for the idea that amenity values for climate variables are capitalized into house prices. Italians would prefer a dryer and sunnier climate during the summer months. But overall the climate of Italy can with some justification be described as optimal as the level of climate variables which maximize land prices is shown to differ insignificantly from sample averages.

Chapter 6 is the generous contribution of Professor Bernard van Praag and Paul Frijters, both of the University of Amsterdam. It measures welfare and well-being in Russia on the basis of two large Russian household surveys carried out in 1993 and 1994. Van Praag and Frijters use the 'income evaluation question' technique to investigate how climatic conditions in various parts of Russia affect the cost of living. This technique invites individuals to state how much income they would require to achieve a standard of living described by a verbal descriptor. These amounts are then analysed to determine the perceived influence of climate on the cost of living. Climatic equivalence scales have been constructed for welfare. The authors find that, as one might expect, the cost of living is negatively affected by cold and harsh winters experienced by some regions and also by high temperatures and humidity.

Chapter 7 has been generously donated by Professor Robert Mendelsohn of Yale University. Once more using the hedonic technique, this chapter quantifies people's preferences to live and work in different climates across the US by examining wage and rent differentials. The study indicates that American people prefer to live in warmer, wetter climates with all their associated, weather and ecosystem characteristics and related and health risks and benefits. The results suggest that the net value of American non-market would increase with mild global warming.

The final chapter provides an important counterpoint to the studies of van Praag and Frijters and Mendelsohn. Using the household production function approach, it examines the climate of India and the extent to which predicted climate changes impact upon the cost of living. Although the data are only cross-sectional in nature it is possible to identify the existence of a climatic optimum, any deviation from which increases the cost of living. The majority of India already exceeds this optimum so that any predicted increase in temperature leads to significant increase in the cost of living in this, the second most populous nation on earth.

Chapter 1

THE AMENITY VALUE OF THE CLIMATE OF BRITAIN

Hedonic theory suggests that if individuals are freely able to select from differentiated localities then the tendency will be for the benefits associated with particular amenities to become collateralized into land prices and wages. Households are attracted to cities offering preferred combinations of environmental amenities and this inward migration both increases land prices within those cities as well as depressing the wage rates in local labour markets (but see below for an important qualification). Thus, across different cities there must generally exist both compensating wage and house price differentials. In such cases both the consumption of amenities and disposable income become choice variables and the value of marginal changes in the level of amenities can be discerned from the hedonic house and wage price regressions. Hedonic theory has a natural application to the determination of the amenity value of climate variables. Combined with predictions concerning the scale and direction of possible climate change, implicit prices of climate variables from regional markets for labour and land might form the basis for a money-metric measure of the resultant change in amenity values to households.

This chapter reports on a study which attempts to determine the extent to which differing climates and the services they provide are collateralized into land prices and wage rates in Britain. The chapter commences with a résumé of the existing hedonic literature concerning the amenity value of climate variables and is followed by a discussion of the theory underlying the hedonic technique. The chapter then discusses the plausibility of some of the assumptions underlying the hedonic technique in the context of uncovering the amenity value of climate variables before describing the data used to estimate the model. Further sections deal with the estimation of the hedonic house and wage price models. The full implicit price of amenities obtained from the analysis is then discussed.

EXISTING LITERATURE

Although there are many early papers describing regressions of land prices on the levels of local amenities (see eg Ridker and Henning, 1967) the theoretical underpinnings of the hedonic approach were developed only later by Rosen (1974). An important theoretical contribution by Roback (1982) demonstrated that values for environmental amenities could be incorporated into both land prices and wage rates. At around the same time others were beginning to consider how to use the hedonic technique to uncover the uncompensated demand curves (rather than just implicit prices) for environmental amenities (eg Brown and Rosen, 1982). Comprehensive reviews of the hedonic approach can be found in Palmquist (1991) and Freeman (1993).

Since Rosen's contribution a large number of hedonic studies have been published in the US. Early studies used either census tract data (ie average house prices in an area) or the self-reported values of those who owned the houses. Later analyses used actual sale price data of individual properties instead. These studies considered both inter-urban and intra-urban regions. Other analyses looked at regional differences in wage rates. The main subject of enquiry in these empirical analyses is the impact of noise from transport on house prices (see Nelson, 1978, for a survey of these) and the impact of air pollution on house prices. Smith and Huang (1991) performed a meta-analysis of 37 different hedonic studies of the impact of air pollution on house prices and concluded that the technique succeeds in providing a consistent set of results; a finding which is generally supportive to the continued use of the hedonic technique.

A considerable number of hedonic studies have included climate variables as environmental amenities. Unfortunately, early studies generally used the hedonic technique primarily to measure the value of amenities such as air quality, crime rates and climate variables using wage data only. They did not consider the possibility of both wages and house prices being affected by the differing availabilities of the environmental amenities (see below). In addition, only one or two climate variables are typically included.

In a seminal paper Hoch and Drake (1974) analysed US wage rates using Bureau of Labor Statistics microdata for 86 Standard Metropolitan Statistical Areas (SMSAs). Climate was specified in terms of precipitation and summer and winter temperature, the squares of the two latter terms, an interaction term for summer temperature and precipitation, wind speed, degree days, snowfall, the number of very hot (>90°F) and the number of very cold (<32°F) days. The analysis was performed separately for three job category subsamples. In the first one, the coefficients on climate variables were significant and had the expected sign only as long as regional dummies were excluded, whereas in the other two they performed better. It must be noted, however, that the only other explanatory variables were racial composition and urban size. No account was given of important site-specific characteristics like crime rates, pollution and other quality-of-life indicators.

The first attempt to estimate the effects of climate on both wages and house prices is to be found in Roback (1982) as an empirical illustration of her theoretical model. She used microdata for the 98 largest urban areas in the US and performed a different pair of regressions (on wages and residential site prices) for each climate variable (snowfall, degree days, cloudy days and clear days). These variables are highly significant and their coefficients have the expected sign in the wages regression, but performed poorly in the residential site prices regressions, where only population growth, population density and unemployment rates are significant. Also for the US, Smith (1983) used only real wages as dependent variables, and employed a cost of living index as a deflator. The focus of the analysis was mainly on the different effects of job-specific and site-specific characteristics on wages for different industries, job categories and ethnic composition of the workers. Climate, expressed by five variables (mean annual sunshine hours, higher and lower temperatures, annual average wind speed and precipitation), was one of the site-specific characteristics considered. Only sunshine proved to have a significant (negative) effect on wages, indicating that it is regarded as an amenity.

Hoehn et al (1987) and Blomquist et al (1988) used the same empirical analysis as empirical illustration of two slightly different theoretical models. Microdata drawn from 285 SMSAs in the US was used. They estimated two hedonic regressions, one for wages and one for housing expenditure. They controlled for structural characteristics of houses and individual characteristics of workers. The same amenity variables were used in both equations and included coast proximity, crime rates, teacher–pupil ratios, total suspended particulates, visibility and six climate variables (sunshine, precipitation, humidity, wind speed, heating and cooling degree days). Amenity variables were found to be highly significant in both equations. Among the climate variables, only sunshine was unambiguously an amenity, that is its coefficient displayed the same sign in both equations, whereas the other climate variables' net effect depended on the relative magnitude of their coefficients. In Blomquist et al (1988) the same set of results was used to derive quality-of-life rankings for the metropolitan areas considered.

Most recently, Maddison and Bigano (1996) analysed property prices per square metre of floor space over 95 provinces of Italy. This analysis is notable in three respects; firstly, it was undertaken specifically to investigate the amenity value of climate variables; secondly, the authors experiment with two alternative specifications of the climate variables; and thirdly, the authors analyse property prices for five consecutive years and find that the amenity values are stable over the period in question. This suggests that the hedonic methodology might be capable of producing figures that are sufficiently reliable for policy purposes (and also that the income elasticity of demand for climate variables is rather low). The authors include three climate variables (temperature, precipitation and cloud cover) and experiment with using annual averages and January and July averages. The climate variables are all highly significant (partly one may suppose because Italy possesses a very diverse climate). The results based upon January and July averages are the best and suggest that the amenity value of additional units of particular climate variables depends upon which season they fall into. In particular, whereas precipitation is viewed as a disamenity in summer or winter, temperature is viewed as an amenity in winter but as something of a disamenity in summer.

MEASURING THE VALUE OF AMENITIES

Roback (1982) provides a formal model in which individuals make decisions in both the market for labour and the market for housing. In her paper the household faces a constraint on disposable income, linking income to the quantity of land occupied for housing (L), non-wage income (I), wages (w) and time spent working (T). Both wages and land prices (r) are a function of the level of the amenities (z) associated with each location. The price of the composite marketed commodity (X) is normalized at unity. The Lagrangian point is:

$$\max U = U(X,T,L,z) + \lambda\left[I + Tw(z) - Lr(z) - X\right] \tag{1}$$

where U is utility. Associated with this problem of constrained maximization is the indirect utility function V containing non-labour income, wage rates, land prices and the level of the environmental amenity as its arguments:

$$V(I,w,r,z) = k \tag{2}$$

Because z is presumed to be an amenity $dV/dz > 0$. In equilibrium the utility associated with each location must be equalized at level k otherwise some individuals would have an incentive to move. Turning to the production side of the economy, assume that a numeraire good is produced by a constant returns to scale technology in which land (L), man-hours of labour (N) and the amenity are arguments:

$$\min TC = wN + rL + \lambda\left[X(N,L,z) \geq \overline{X}\right] \tag{3}$$

Associated with this technology is a unit cost function C. The equilibrium condition is that the unit cost function must equal a constant. If it did not then some firms would have an incentive to move to a different location.

$$C(w,r,z) = 1 \tag{4}$$

Totally differentiating the two equilibrium conditions gives:

$$\frac{\partial V}{\partial w}\frac{dw}{dz} + \frac{\partial V}{\partial r}\frac{dr}{dz} + \frac{\partial V}{\partial z} = 0 \tag{5}$$

$$\frac{\partial C}{\partial w}\frac{dw}{dz} + \frac{\partial C}{\partial r}\frac{dr}{dz} + \frac{\partial C}{\partial z} = 0 \tag{6}$$

Solving for the expressions dw/dz and dr/dz gives:

$$\frac{dw}{dz} = \frac{-V_z C_r + C_z V_r}{V_w C_r - V_r C_w} \tag{7}$$

and:

$$\frac{dr}{dz} = \frac{-V_w C_z + C_w V_z}{V_w C_r - V_r C_w}$$

8

Given the signs of the various derivatives it can be seen that the following conditions hold: if the amenity is unproductive (ie $C_z > 0$) or is neutral with respect to production costs, then the wage gradient is negative and the rent gradient is positive. But if the amenity is productive then the sign of the wage gradient is ambiguous since workers may not be forced to accept lower wages by the ability of the firm to relocate. If the amenity is neutral or productive to firms then the rent gradient is positive. But if the amenity is unproductive then the rent gradient is ambiguous since land prices might need to fall in order to encourage firms to locate in areas with high levels of z.

Using Roy's Identity theorem and solving for the implicit price of the amenity, it can be shown that the implicit price of the amenity to the household is the sum of the amount of land occupied by the household multiplied by the marginal cost of obtaining land, with an additional unit of the amenity minus the marginal change in labour income associated with working in an area with an additional unit of the amenity. In general, both the changes in land prices and wage rates have to be considered in order to determine the implicit price of the amenity to the household.

$$p_z = L\frac{dr}{dz} - T\frac{dw}{dz}$$

9

The amount of land consumed by each household, the hours of work, dw/dz and dr/dz are in principle observable entities as are the amount of land occupied by each producer and the share of land and labour in overall production costs.

Additional tradeable goods sectors and additional amenities add nothing new to the story. More interesting extensions of the model by Roback include the addition of a non-traded goods sector and a second set of workers. The introduction of a non-tradeable goods sector modifies the conclusions of the model slightly. Roback (op cit) assumes that there is a single non-tradeable goods sector producing services. This sector also enjoys constant returns to scale and has a unit cost function which can be written:

$$G(w,r,z) = p(z)$$

10

Totally differentiating this expression gives:

$$\frac{\partial G}{\partial w}\frac{dw}{dz} + \frac{\partial G}{\partial r}\frac{dr}{dz} + \frac{\partial G}{\partial r} = \frac{dp}{dz}$$

11

The household's indirect utility function must also be amended to include the price of non-tradeables and is written:

$$V(I,w,r,p,z) = k$$

12

Totally differentiating this expression then gives:

$$\frac{\partial V}{\partial w}\frac{dw}{dz} + \frac{\partial V}{\partial r}\frac{dr}{dz} + \frac{\partial V}{\partial p}\frac{dp}{dz} + \frac{\partial V}{\partial z} = 0 \qquad\qquad 13$$

The three equations (6), (11) and (13) can be solved to yield expressions for dw/dz, dr/dz and dp/dz:

$$\frac{dw}{dz} = \frac{C_z(V_r + V_p G_r) - C_r(V_z + V_p G_z)}{C_r(V_w + V_p G_w) - C_w(V_r + V_p G_r)} \qquad\qquad 14$$

$$\frac{dr}{dz} = \frac{C_z(V_w + V_p G_w) - C_w(V_z + V_p G_z)}{C_w(V_r + V_p G_r) - C_r(V_w + V_p G_w)} \qquad\qquad 15$$

$$\frac{dp}{dz} = G_w \frac{C_z(V_r + V_p G_r) - C_r(V_z + V_p G_z)}{C_r(V_w + V_p G_w) - C_w(V_r + V_p G_r)} +$$

$$G_r \frac{C_z(V_w + V_p G_w) - C_w(V_z + V_p G_z)}{C_w(V_r + V_p G_r) - C_r(V_w + V_p G_w)} + G_z \qquad\qquad 16$$

The implicit price of the amenity is once again derived by the use of Roy's theorem to obtain:

$$p_z = L\frac{dr}{dz} - T\frac{dw}{dz} + G\frac{dp}{dz} \qquad\qquad 17$$

where G is the quantity of non-tradeable goods purchased (again G could be a function of z). This makes it clear that when non-tradeable goods are brought into the picture, hedonic price schedules also need to be calculated for them. The sign of dp/dz is generally ambiguous (even in the case where the amenity is unproductive to both the traded and the non-traded good). But while theory suggests that hedonic regressions are required for local service price indices (excluding housing) such detailed data do not exist at the county level in Britain. The empirical implementation of the model described below therefore abstracts from reality in that it assumes that all goods are tradeable.[1]

A CRITIQUE OF THE HEDONIC APPROACH

There are several acknowledged problems with the hedonic method and the assumptions which underpin it. These relate to problems of individual perception, subjectivity, continuity, averting behaviour, market segmentation and the assumption of equilibrium: they need to be addressed prior to confronting the model just outlined with real-world data.

Individual perceptions are important because amenity values will only be reflected in house and wage price differentials to the extent that individuals are aware of differences in amenity levels and the effect that these might have on health, for example (Freeman, 1993). Lack of information is a criticism which is often levelled against the use of hedonic studies to determine the amenity value of air quality where individuals are subjected to some pollutant, the health effects of which they are unaware.

It is important in hedonic analyses to include all relevant variables, as variable omission can lead to biased estimates. But there is no accepted list of which environmental variables need to be controlled for. Furthermore climate in particular can be described in a large number of ways. One possible response might be to include all possible variables from the outset, but the inclusion of irrelevant variables leads to increased variance in the estimates. Furthermore with environmental variables problems of multicollinearity frequently arise (Freeman, 1993). It is clearly difficult to measure the individual contribution of particular variables to overall amenity levels.

To the extent that 'averting' behaviour is possible, this should be accounted for in the hedonic equation. Averting behaviour refers to the purchase of goods partly or wholly for reasons connected with reducing the direct effect of environmental disamenities on utility. To understand why averting behaviour might be an issue in the context of hedonic studies into the amenity value of climate, in theory what is required is a measure of variations in land prices with levels of environmental variables. But seldom are the financial details of transactions in land made public and most studies resort to using house prices instead. This is satisfactory insofar as building land is not generally sold separately from the house which stands upon it. However, certain characteristics of houses (such as double glazing, loft insulation and central heating) are likely to be correlated with climate variables, while simultaneously contributing to the value of the property and thus imparting bias to the coefficients on the variables of interest. If the extent of these averting expenditures were known, they could be included as a separate regressor in the hedonic house price function. Partly for this reason there is a general presumption in favour of using data detailing actual market transactions concerning individual properties with known characteristics rather than census tract data (ie data dealing in average prices rather than the price of particular properties). On the other hand individual observations contain a large amount of 'noise' which can be overcome only by including a large number of observations and/or by controlling for a range of variables which are not the main focus of the study (Palmquist, 1991).

The theory underlying the model relies on the existence of smoothly continuous trade-off possibilities among all characteristics. In other words, all possible combinations of housing characteristics should be available on the market. This is necessary for households to be able to locate at a position of simultaneous equilibrium with respect to all characteristics. Without this assumption the observed implicit price of the amenity cannot be taken as an estimate of marginal willingness to pay for it, since the household may be located at a 'corner solution'. This assumption seems to be a reasonable one in the context of climate variables which vary smoothly over adjacent areas (Maler, 1977).

If there is restricted mobility among different sections of an area, then different markets will exist and pooling of data might lead to bias since only a single regression line is effectively fitted to two or more spline functions

(Straszheim, 1974). Restricted mobility might well present a problem to hedonic studies into the amenity value of climate, since climate variables only vary significantly over relatively large distances, at which points the assumption of unrestricted mobility becomes harder to defend because of cultural and even language differences. Fortunately, the stability of the hedonic price functions across different regions is a testable hypothesis. Finding that the hedonic price function is not stable across different regions means that the separate hedonic price functions must be fitted to each, but it does not render the hedonic technique invalid.

Finally, there is the heroic assumption of perfect equilibrium in the housing market and the labour market. For this assumption to hold there must be perfect information, zero transaction costs, zero moving costs and perfectly flexible prices. Evans (1990) argues that the assumption of equilibrium means that the environmental quality values derived from hedonic analyses are likely to be systematically incorrect and should not without qualification be used as measures of attractiveness of an area. However, if climate amenities are valued, then in the process of equilibration following supply or demand side shocks migration ought to be observed flowing in the direction of more pleasant climates and away from regions where the climate is less pleasant.[2] In the approach favoured by 'disequilibrium' theorists, net or gross migration rates are regressed on a set of regional variables to establish the trade-off between financial variables and environmental amenities (eg Greenwood and Hunt, 1986).

But the conclusion that amenity values for environmental variables derived from hedonic analyses are likely to be systematically incorrect is wrong. There is no reason to suppose that the implicit prices derived from hedonic analyses are biased because there is no *a priori* reason to suppose that the extent of disequilibrium in any area is correlated with the levels of particular amenities. The consequence of disequilibrium is likely to be an increased variance in results rather than systematic bias (Freeman, 1993). In fact the debate between the proponents of the 'equilibrium' and 'disequilibrium' approaches actually involves a matter of emphasis on different aspects of the location decision (Hunt, 1993). The different positions are characterized by beliefs concerning the speed of adjustment and the motivation for migration, for example. The disequilibrium approach reviewed in Greenwood (1985) sees adjustment to equilibrium as being extremely slow and equilibrium concepts of little relevance. The equilibrium approach typified by the work of Blomquist et al (1988) sees local labour markets and housing markets as operating relatively efficiently. Disequilibrium theorists see migration primarily as a response to real utility differences between locations whereas equilibrium theorists tend to emphasize changes in the level of consumption amenities and life-cycle events in migration. Were it the case that both of these approaches could be based explicitly on utility maximization, which of them constitutes the superior approach for the purposes of obtaining monetary estimates of the amenity value of climate variables would be a purely empirical matter. However, unlike the hedonic technique, the migration-based approach is not typically based on utility maximization.

DO HOUSE PRICES REFLECT DIFFERENCES IN THE LEVEL OF CLIMATE AMENITIES?

The chapter now turns to consider the estimation of the first component of the implicit price of climate: the hedonic land price schedule. This study uses census tract property prices relating to 127 English and Welsh counties, Scottish regions, metropolitan areas and London boroughs as the dependent variable instead of difficult-to-obtain data on land prices.[3] The census tract data are taken from Focas et al (1995) and refer to the year 1994. The prices correspond to a straight average across five different property types (terraced house, semi-detached house, detached house, bungalow, flat/maisonette). Since outright purchase prices are used, rather than annual rental values, the implicit prices which emerge reflect the discounted stream of benefits over the remaining lifetime of the house.

Turning to the explanatory variables, an oft-encountered problem involved in the use of the hedonic technique is how best to represent a fluctuating amenity such as the climate in the regression. Normal practice is to enter the annual mean as a summary statistic. But there is evidence that using information on the distribution of the level of the amenity can significantly increase the statistical performance of the hedonic price regression. One possibility therefore might be to include a variable representing the expected number of days per year when temperatures exceed 90°F or remain below freezing. The number of days on which a gale blows might also be a far more useful measure of amenity rather than average windspeed, if indeed households are primarily concerned about extremes of weather rather than averages. It might also be useful to experiment with the concept of heating and cooling degree days (the annual average cumulative deviation on either side of 65°F). Heating and cooling degree days have regularly been used with some success in American studies as indices of climate (see eg Blomquist et al, 1988).

The only published attempt to discriminate between different specifications of climate variables of any sort is Cushing (1987). Cushing's analysis deals only with temperature variables and uses non-nested tests to examine the power of alternative specifications of temperature variables in the context of an inter-state study of migration in the US. He examines the concept of average temperatures, heating and cooling degree days and average temperatures during the hottest and coldest months. His findings suggest that the latter concept is the single most appropriate one whilst annual average temperatures are rather poor. Cushing rationalizes this by stating that the annual averages cannot distinguish between a climate that is mild all the year round and one which is extremely cold in winter and extremely hot in summer. Similarly the concept of heating and cooling degree days does not distinguish between deviations above or below an arbitrary point (typically 65°F). Cushing's findings should be seen as referring primarily to the case of the US for which the concept of heating and cooling degree days may be particularly apt. This is because the US can be described as having a 'continental' climate (ie one of seasonal extremes). Britain, however, enjoys what could be described as a 'maritime' climate and this study restricts itself to examining the impact of average climates.

Annual average values of climate variables (1961–1990) were available on a 10 km grid square basis measured at the average altitude within each grid square.[4] A gazetteer (Ordnance Survey, 1992) of Great Britain was used to determine the grid reference location of the major conurbations. From this procedure average values were obtained for precipitation, temperature and sunshine for each county, region or borough while avoiding uninhabited regions. Values for humidity levels and average windspeed were not included in the analysis: humidity is only important in either very hot or very cold conditions, neither of which apply to Britain, and average windspeed (unlike the number of gale-days) is an unfamiliar concept. Apart from climate variables a range of other amenity variables are also controlled for. These include the local crime rate, local taxes, school quality variables, the quality of health care services, transport links, population density and unemployment. These are now discussed in turn.

Crime statistics in Great Britain are recorded at an individual police force level by the Home Office and the Scottish Office (see Focas et al, 1995, for precise details). One could make a case for using several alternative measures of criminal behaviour rather than just using the rate for all reported crime, but it was decided to select one from a list of three possible measures of criminal activity: all crime, violent crime and burglaries per thousand of population. Of these alternatives the number of burglaries appeared to provide the best explanation of variations in house prices and wage rates. The frequency of burglaries presumably increases the cost of residing in a given area insofar as it raises the cost of home insurance.

The importance of including variables describing local services and fiscal conditions is discussed in Gyourko and Tracy (1989) and Charney (1993). Both argue that what is required are separate measures of local taxes and the quality of local services rather than measures of expenditure on local services. The problem with employing expenditure on local services is that expenditure on its own cannot distinguish between the existence of a high quality service and a service which is expensive because it is managed ineptly. Furthermore, Charney argues that higher local taxes are a disamenity only insofar as they do not reflect higher quality or more extensive provision of public services. If these factors are not adequately controlled for then it is possible that the implicit price of additional taxation could be zero or even positive. Local 'council' taxes are set on a sub-county level and 'Band D' tax levels are averaged to yield a single figure for each county.

It is of course very difficult to find suitable indices to measure the quality of local services and the best that can be done is to use the often very crude proxies that are available at the correct level of regional aggregation. In the case of education services, the most widely cited measure of quality relates to the percentage of students obtaining five General Certificate of Secondary Education (GCSE) passes at Grade C or better (alternative measures used in other studies include the number of students per teacher or the rate of truancy). In the case of health services, the proxy used for the quality of the service is the number of patients per General Practitioner (GP) (an alternative might be the length of waiting lists for key surgical procedures, although such a variable is only available at the level of regional health authorities rather than counties). These statistics are available from the Department of Education and the Department of Health respectively (again see Focas et al, 1995, for details). The abundance of transport links is crudely approximated by the number of railway stations per unit area multiplied by 100.

There are many other factors besides climate which determine the level of environmental quality to be enjoyed in a particular area. These include the ambient concentration of air pollution in a particular locality (see eg Smith and Huang, 1991), the level of noise nuisance (Nelson, 1978) and the extent of local traffic congestion (Maddison and Bigano, 1996). There might even be benefits in the form of economies of agglomeration. All these phenomena are a reflection of physical proximity to one's neighbours and, in the absence of more reliable measures, are proxied by population density.

The definition of population density employed in the empirical model is quite different from the definition of population density in the theoretical model. The latter is simply the inverse of land for habitation per individual. The former concept, population per unit of all-land area, is the more commonly employed. While these measures are not entirely unrelated, the majority of hedonic studies treat all-land population density as an exogenous regressor.[5] Population density is taken from the Central Statistical Office (1995).

Unemployment has been frequently included as a disamenity in hedonic house price and wage rate analyses. One can interpret unemployment as a disamenity in the sense that households require compensation to live in areas characterized by scant opportunities for employment (Todaro, 1969, and Harris and Todaro, 1970). In the context of this latter explanation one would expect the full implicit price of an additional percentage point on the local unemployment rate to be negative. But instead of treating unemployment as a disamenity a more realistic formulation of the locational problem is one in which the expected utility of all locations is equalized, given that there are two possible states: employed and unemployed. However, reformulating the problem would not change the analysis except in a trivial way. In particular, one would still expect to find differences in unemployment rates reflected in local land prices and wage rates.[6]

Outside the hedonic literature there have been numerous empirical analyses of the influence of unemployment on regional labour markets. Savouri (1989), for example, considers ten regional labour markets in Britain over time and finds that local unemployment exerts strong downward pressure on regional wage rates. Unemployment rates are from the Department of Employment's *New Earnings Survey* (DOE, 1995).

Finally, three dummy variables are used to test for (and at the same time hopefully correct for) the existence of a segmented market for housing. A separate dummy variable is used if the observation is drawn from a London borough, Scotland or Wales. This allows the intercept of the hedonic price regression to vary across the different regions (common slopes across the different regions is a maintained hypothesis because of the degrees of freedom constraint). The inclusion of dummy variables for Scotland and Wales has the unfortunate effect of identifying respectively the coldest and wettest parts of Britain, and in this sense they clearly compete with the climate variables.

The variables included in the hedonic house price regression are shown in Table 1.1 and the range of values that these variables take is shown together with their mean values and standard deviations in Table 1.2. From inspecting the mean to standard error ratio it is apparent that whereas precipitation varies markedly across Britain, the other climate variables do not. It is therefore likely that among the climate variables only the coefficient on rainfall will be determined with any real degree of precision.

Table 1.1 *Definition of Variables included in the Hedonic Land Price Regression*

Variable	Definition
HOUSE	Current purchase price of property in 1994 (£)
BURGLARY	Number of reported burglaries per 1000 of population (1993)
TAX	Council tax at Band D (£)
RAIL	Railway accessibility (number of railway stations divided by area multiplied by 100)
GP	Average number of patients per doctor (1994)
EXAM	Percentage of students obtaining 5 GCSE exam passes (Grade C or better) or Scottish equivalent
POPDEN	Population density (persons per square kilometre)
UNEMP	Percentage of economically active workforce unemployed
PRECIP	1960–1991 average precipitation (mm)
TEMP	1960–1991 average temperature (°C)
SUN	1960–1991 average hours of sunshine
LONDON	Dummy variable which takes the value unity for Greater London, zero otherwise
SCOTLAND	Dummy variable which takes the value unity for Scotland, zero otherwise
WALES	Dummy variable which takes the value unity for Wales, zero otherwise

ESTIMATION

Once the dependent variables have been selected, one is left with the task of determining the appropriate functional form. It is typically found that fitting an inappropriate functional form to hedonic price equations has serious implications for the implicit prices which emerge (Palmquist, 1991). The most desirable procedure is to employ transformations which are flexible enough to accommodate a variety of functional forms. Perhaps the most rigorous procedure is to apply the Box and Cox (1962) transformation to all of the dependent and independent variables as suggested by Halvorsen and Pollakowski (1981). For the purposes of this study, however, such a procedure is unfeasible because of a degrees-of-freedom constraint. The one concession made to the functional form of the hedonic house price

Table 1.2 *The Characteristics of the House Price Data Set*

Variable	Mean	Standard deviation	Minimum	Maximum
HOUSE	65640	22462	0.4199E+05	0.1852E+06
BURGLARY	26.063	11.325	1.000	55.00
RAIL	14.113	25.270	0.0000	139.2
TAX	450.05	71.323	256.8	657.8
GP	1885.8	220.28	1131.	2219.
EXAM	38.205	7.9330	18.00	53.00
POPDEN	2168.6	2714.6	8.000	0.1247E+05
UNEMP	8.7449	3.2675	2.900	22.40
PRECIP	789.98	210.33	568.0	1640
TEMP	9.2472	0.84310	7.000	10.60
SUN	1404.2	118.32	1044	1770.
LONDON	0.25197	0.43457	0.0000	1.000
SCOTLAND	0.94488E-01	0.29280	0.0000	1.000
WALES	0.62992E-01	0.24319	0.0000	1.000

regression was to attempt the transformation of the dependent and independent variables considering four frequently discussed cases:

$$\frac{HOUSE_i^\lambda - 1}{\lambda} = \alpha + \beta_1 \frac{BURGLARY_i^\theta - 1}{\theta} + \beta_2 \frac{RAIL_i^\theta - 1}{\theta} +$$

$$\beta_3 \frac{TAX_i^\theta - 1}{\theta} + \beta_4 \frac{GP_i^\theta - 1}{\theta} + \beta_5 \frac{EXAM_i^\theta - 1}{\theta} + \beta_6 \frac{POPDEN_i^\theta - 1}{\theta} +$$

$$\beta_7 \frac{UNEMP_i^\theta - 1}{\theta} + \beta_8 \frac{PRECIP_i^\theta - 1}{\theta} + \beta_9 \frac{TEMP_i^\theta - 1}{\theta} +$$

$$\beta_{10} \frac{SUN_i^\theta - 1}{\theta} + \beta_{11} LONDON + \beta_{12} SCOTLAND + \beta_{13} WALES + e_i \qquad 18$$

The four cases are: $\lambda = 0,1$ and $\theta = 0,1$. In the log-log, semi-log and lin-log model, the marginal value of any one characteristic depends upon the value of all other characteristics. Only in the linear model is the marginal value of a change in any one of the characteristics independent of the level of any other characteristic. Note that the dummy variables LONDON, SCOTLAND and WALES are not transformed.

RESULTS

Using the method described by Maddala (1977) it was found that a semi-logarithmic model (corresponding to the case where $\lambda = 0$ and $\theta = 1$) was the model more likely to have generated the observed data and this specification was adopted for the remainder of the study. The estimated coefficients of the semi-logarithmic model are shown in Table 1.3. The regression analysis

Table 1.3 *The Estimated Hedonic House Price Regression*
(Semi-Logarithmic Model)

Ordinary least squares regression			Dependent variable =		Log HOUSE
Observations	=	127	Weights	=	ONE
Mean of			Standard		
left hand side	=	0.1104822E+02	deviation of		
			left hand side	=	0.2796593E+00
Standard			Sum of squares	=	0.1925348E+01
deviation of					
residuals	=	0.1305315E+00			
R-squared	=	0.8046199E+00	Adjusted R-squared =		0.7821426E+00
F[13, 113]	=	0.3579692E+02	Probability value		0.0000000E+00
Log-likelihood	=	0.8580140E+02	Restriction		
			($\beta=0$) Log-l	=	−0.1788195E+02
Amemiya			Akaike		
probability			information		
criterion	=	−0.1130731E+01	criterion	=	0.1891674E−01

ANOVA source	Variation	Degrees of freedom	Mean square
Regression	0.7929026E+01	13	0.6099251E+00
Residual	0.1925348E+01	113	0.1703848E−01
Total	0.9854374E+01	126	0.7820932E−01

Variable	Coefficient	Standard error	t-ratio	Probability x
Constant	9.8870	0.4555	21.706	0.00000
BURGLARY	−0.29835E−02	0.1512E−02	−1.973	0.05098
RAIL	0.43612E−02	0.1005E−02	4.340	0.00003
TAX	0.95164E−03	0.2391E−03	3.980	0.00012
GP	−0.36255E−04	0.1074E−03	−0.338	0.73630
EXAM	0.16801E−02	0.3268E−02	0.514	0.60813
POPDEN	0.55751E−04	0.1275E−04	4.374	0.00003
UNEMP	−0.49220E−01	0.7714E−02	−6.381	0.00000
PRECIP	−0.15642E−03	0.7761E−04	−2.016	0.04622
TEMP	0.90885E−01	0.4230E−01	2.148	0.03381
SUN	0.24155E−03	0.1937E−03	1.247	0.21492
LONDON	−0.80640E−01	0.5847E−01	−1.379	0.17059
SCOTLAND	0.23892	0.8475E−01	2.819	0.00569
WALES	0.71696E−01	0.6426E−01	1.116	0.26691

manages to explain in excess of 80 per cent of the variation in house prices; a fact which reflects the use of census tract data and the fact that the characteristics of individual properties are averaged out of the data. That amenity variables are able to explain differences in regional house prices suggests that the 'equilibrium' (hedonic) approach to amenity values is the appropriate one. Property prices evidently contain a great deal of information on the value of amenities and cannot therefore be 'too far' away from equilibrium.

Dealing first with the non-climate variables, it is apparent that while some of these play a highly significant role in determining residential property prices, others may fail to capture the quality of local services. To begin with, the variable describing the rate of burglaries appears to significantly affect property prices in the anticipated direction. The number of transport terminuses is also a statistically significant determinant of property prices. But on the other hand, the tax variable is highly significant and unexpectedly signed. Possibly this is because, as indicated earlier, this variable plays a dual role in acting as a proxy for higher levels of public services as well as indicating a higher charge to pay. The unexpectedly signed coefficient on this variable can thus be interpreted as a failure to control adequately for the differences in the levels of public services paid for through local taxes. This interpretation is borne out by the fact that the variables describing the quality of health care services and education services are statistically insignificant. Population density is highly significant and positively signed. Whereas one might have expected it to play a role as a proxy for various forms of pollution such as noise pollution, air pollution and congestion, these factors seem to be outweighed by benefits from agglomeration. Local unemployment, on the other hand, reduces house prices markedly.

Turning to the climate variables, it is evident that a house in a warm and sunny location is significantly more expensive than a house in a wet location. An F-test of the joint significance of the climate variables was performed and confirmed that the hypothesis that the coefficients on these three variables were all simultaneously zero could be rejected even at the 99 per cent level of confidence.[7] Thus this study provides very strong support in favour of the hypothesis that amenity values for climate variables are embedded within British property prices. It is also interesting to note that one of the dummy variables, that for Scotland, is significant, suggesting that the housing market is segmented in Britain along a north-south axis.

Do Wages Compensate for an Unpleasant Climate?

Having shown that amenity values for climate variables are to a significant extent embedded in property prices, the following sections seek to determine whether differing levels of environmental amenities, and in particular climate amenities, are similarly reflected in regional wage differences across Britain. It was earlier shown that the only circumstances under which the land price gradient would capture the entirety of the implicit price of environmental amenities would be in a situation where firms do not use land in production and production is unaffected by the level of the amenity. In these circumstances wages cannot differ from location to location because this would necessarily imply different production costs which competition would eventually eliminate. It is not even certain that the hedonic wage gradient need be negative with respect to the level of an environmental amenity.

REGIONAL WAGE DATA

Pooled average hourly wage rates for full-time workers relating to 127 English and Welsh counties, Scottish regions, metropolitan areas and London boroughs are analysed. These are taken from the *New Earnings Survey* (DOE, 1995) and refer to the year 1994. The characteristics of the regional wage data are described in Table 1.4. The hourly wage rates exclude overtime payments and consider only full-time workers. The study is hampered by an inability to standardize for many of the characteristics of different workers or the characteristics of employment. In fact, only two characteristics of the workers can be controlled for: differences between manual and non-manual labour and differences between male and female workers. Remaining differences between workers (eg educational attainment and racial composition) and characteristics of the industry or the job (eg the extent of unionization) are consigned to the error term. Neither industry-based wage rates nor micro-level data detailing the characteristics of individual workers and their employment are available at the required level of disaggregation for Britain (ie by county).

Besides worker characteristics, the same set of amenities as was used in the hedonic house price regression is entered into the hedonic wage regressions: measures of criminal activity, the level of local taxes, transport linkages, the quality of health care, the quality of local schools, population density, unemployment, climate variables and dummy variables for different regions to test for possible segmentation of the labour market.

Table 1.4 *The Characteristics of the Regional Wage Data Set for 1994 (pence per hour)*

Category	Mean	Standard deviation	Minimum	Maximum
MAN. MALE	627.02	56.470	513.0	841.0
MAN. FEMALE	448.34	29.156	395.0	521.0
NON MAN. MALE	1079.2	163.11	880.0	1620.0
NON-MAN. FEMALE	736.83	107.06	551.0	1115.0

Source Department of Employment (1995)

ESTIMATION AND RESULTS OF THE HEDONIC WAGE REGRESSION

The first model estimated was a 'pooled' model which combines the four different categories of labour and uses dummy variables to control for differences in the intercept of the hedonic wage regression between male, female, manual and non-manual workers. The one concession made to functional form is to attempt the transformation of the dependent and independent variables by the means suggested by Box and Cox (1962). The following model was considered (note that the dummy variables are not transformed):

$$\frac{WAGE_i^{\lambda} - 1}{\lambda} = \alpha + \beta_1 \frac{BURGLARY_i^{\theta} - 1}{\theta} + \beta_2 \frac{RAIL_i^{\theta} - 1}{\theta} +$$

$$\beta_3 \frac{TAX_i^{\theta} - 1}{\theta} + \beta_4 \frac{GP_i^{\theta} - 1}{\theta} + \beta_5 \frac{EXAM_i^{\theta} - 1}{\theta} + \beta_6 \frac{POPDEN_i^{\theta} - 1}{\theta} +$$

$$\beta_7 \frac{UNEMP_i^{\theta} - 1}{\theta} + \beta_8 \frac{PRECIP_i^{\theta} - 1}{\theta} + \beta_9 \frac{TEMP_i^{\theta} - 1}{\theta} +$$

$$\beta_{10} \frac{SUN_i^{\theta} - 1}{\theta} + \beta_{11}LONDON + \beta_{12}SCOTLAND + \beta_{13}WALES +$$

$$e_i + \beta_{14}MALE + \beta_{15}MANUAL + e_i \qquad\qquad 19$$

Once again four special cases are entertained: $\lambda = 0,1$ and $\theta = 0,1$ and once more it was found that the combination $\lambda = 0$ and $\theta = 1$ was most likely to have generated the observed data. The pooled model, however, constrains the slopes of the regression equation to be identical across the different worker categories. Previous researchers have run separate regressions for different occupations and skill groups, arguing that since different groups of workers compete in different markets there is no reason why the slopes (and hence the implicit prices paid by workers of each type) should be the same (Roback, 1988).[8] Alternatively, it may be believed that the underlying assumptions of the hedonic technique are more likely to be met for one group of workers rather than the other. More specifically, it might be that greater mobility of skilled workers means that only they are able to respond to real utility differentials existing between locations. If this were so then climate variables might be significant in the hedonic wage regressions for skilled workers but not in those for unskilled workers.

In order to test the hypothesis of common slopes, a separate regression is run for each category of worker (male/female/manual/non-manual) and the sum of squared residuals compared with the sum obtained from the pooled model. The results of this F-test suggest that the slopes of the model do indeed differ substantially across the four groups.[9] The estimated regression equations for each category of worker are presented in Tables 1.5 to 1.8. Next, a test of the joint significance of the coefficients of the climate variables is performed in each of the worker-specific hedonic wage equations. In three out of the four categories the hypothesis of zero slopes on the climate variables cannot be rejected at the 95 per cent level of confidence. But in the case of non-manual female workers the hypothesis of zero slopes on the climate variables can be rejected at this level of confidence (although not at the 99 per cent level).[10] The evidence therefore appears to be rather against the compensation for the amenity value of climate variables through regional differences in wage rates, at least in Britain.

Turning to the unrestricted regressions, their ability to explain variations in regional wage rates varies between 64 per cent and 82 per cent of the total variation around the mean. The significance of the dummy variable LONDON points to the existence of a segmented labour market. As expected, the coefficient on unemployment is either negative and significant (or in one case positive and insignificant). There are no other consistent patterns regarding the significance of the remaining amenity variables between the respective regressions. In particular, *none* of the climate variables is individually significant at 95 per cent level of confidence in *any* of the regressions.

Table 1.5 *The Estimated Hedonic Wage Regression: Male Manual Workers*

Ordinary least squares regression			Dependent variable =		Log WAGE
Observations	=	127	Weights	=	ONE
Mean of left hand side	=	0.6437117E+01	Standard deviation of left hand side	=	0.8752555E-01
Standard deviation of residuals	=	0.5489917E-01	Sum of squares	=	0.3405728E+00
R-squared	=	0.6471666E+00	Adjusted R-squared	=	0.6065751E+00
F[13, 113]	=	0.1594342E+02	Probability value		0.0000000E+00
Log-likelihood	=	0.1957982E+03	Restriction (ß=0)		
			Log-l	=	0.1296465E+03
Amemiya probability criterion	=	−0.2862964E+01	Akaike information criterion	=	0.3346162E-02

ANOVA source	Variation	Degrees of freedom	Mean square
Regression	0.6246782E+00	13.	0.4805217E-01
Residual	0.3405728E+00	113.	0.3013919E-02
Total	0.9652510E+00	126.	0.7660722E-02

Variable	Coefficient	Standard error	t-ratio	Probability x
Constant	6.1834	0.1916	32.277	0.00000
BURGLARY	−0.11104E-02	0.6361E-03	−1.746	0.08361
RAIL	0.13870E-03	0.4226E-03	0.328	0.74336
TAX	0.16348E-03	0.1006E-03	1.626	0.10682
GP	0.10094E-03	0.4517E-04	2.235	0.02740
EXAM	0.12010E-03	0.1374E-02	0.087	0.93052
POPDEN	−0.30802E-06	0.5360E-05	−0.057	0.95428
UNEMP	0.58772E-02	0.3244E-02	1.812	0.07271
PRECIP	−0.35353E-04	0.3264E-04	−1.083	0.28106
TEMP	−0.45679E-02	0.1779E-01	−0.257	0.79784
SUN	0.26666E-05	0.8146E-04	0.033	0.97394
LONDON	0.10128	0.2459E-01	4.118	0.00007
WALES	0.10438E-01	0.2703E-01	0.386	0.70006
SCOTLAND	0.18238E-01	0.3564E-01	0.512	0.60988

THE FULL IMPLICIT PRICE OF AMENITIES

Having investigated the evidence for the collateralization of amenity values for climate in the hedonic house price and hedonic wage rate regressions, it is now possible to comment on the full implicit price for climate amenities. Although it is the implicit price of climate rather than non-climate amenities which is the main focus of this chapter, it is nevertheless of interest to present

Table 1.6 *The Estimated Hedonic Wage Regression:*
Male Non-Manual Workers

Ordinary least squares regression			Dependent variable =		Log WAGE
Observations	=	127	Weights	=	ONE
Mean of left			Standard deviation		
hand side	=	0.6973714E+01	of left hand side	=	0.1407272E+00
Standard deviation			Sum of squares	=	0.6291774E+00
of residuals	=	0.7461864E-01			
R-squared	=	0.7478573E+00	Adjusted R-squared =		0.7188498E+00
F[13, 113]	=	0.2578146E+02	Probability value		0.0000000E+00
Log-likelihood	=	0.1568229E+03	Restriction (ß=0)		
			Log-l	=	0.6933513E+02
Amemiya			Akaike		
probability			information		
criterion	=	−0.2249180E+01	criterion	=	0.6181730E-02

ANOVA source	Variation	Degrees of freedom	Mean square
Regression	0.1866146E+01	13.	0.1435497E+00
Residual	0.6291774E+00	113.	0.5567942E-02
Total	0.2495323E+01	126.	0.1980415E-01

Variable	Coefficient	Standard error	t-ratio	Probability x
Constant	7.1019	0.2604	27.274	0.00000
BURGLARY	−0.40924E-03	0.8646E-03	−0.473	0.63690
RAIL	0.11721E-02	0.5744E-03	2.041	0.04363
TAX	0.17125E-03	0.1367E-03	1.253	0.21286
GP	−0.87145E-04	0.6139E-04	−1.419	0.15852
EXAM	−0.19039E-02	0.1868E-02	−1.019	0.31026
POPDEN	0.20513E-04	0.7286E-05	2.815	0.00575
UNEMP	−0.13088E-01	0.4410E-02	−2.968	0.00366
PRECIP	−0.40571E-04	0.4436E-04	−0.914	0.36240
TEMP	−0.16719E-01	0.2418E-01	−0.691	0.49074
SUN	0.17622E-03	0.1107E-03	1.592	0.11426
LONDON	0.14351	0.3343E-01	4.293	0.00004
WALES	−0.26813E-01	0.3673E-01	−0.730	0.46696
SCOTLAND	0.68384E-02	0.4845E-01	0.141	0.88800

the implicit prices of non-climate variables since the plausibility or implausibility of those results also has some bearing on the overall credibility of the technique being used.

Using the estimated parameters of the hedonic house price equation and the model in which wage rates for manual and non-manual workers are pooled, the implicit prices of amenities are calculated for the typical household receiving £11,830 annually after tax in wages and salaries in 1994 (CSO, 1995) and a typical property costing £65,640 whose benefits are annuitized using a conventional 5 per cent rate of discount.

The full implicit prices are obtained by subtracting column two from column one in Table 1.9. This reflects the fact that an amenity with a negative

Table 1.7 *The Estimated Hedonic Wage Regression: Female Manual Workers*

Ordinary least squares regression			Dependent variable =		Log WAGE
Observations	=	127	Weights	=	ONE
Mean of left hand side	=	0.6103480E+01	Standard deviation of left hand side	=	0.6436843E-01
Standard deviation of residuals	=	0.4078658E-01	Sum of squares	=	0.1879806E+00
R-squared	=	0.6399219E+00	Adjusted R-squared =		0.5984970E+00
F[13, 113]	=	0.1544776E+02	Probability value		0.0000000E+00
Log-likelihood	=	0.2335356E+03	Restriction (ß=0)		
			Log-l	=	0.1686745E+03
Amemiya probability criterion	=	−0.3457254E+01	Akaike information criterion	=	0.1846928E-02

ANOVA source	Variation	Degrees of freedom	Mean square
Regression	0.3340746E+00	13.	0.2569804E-01
Residual	0.1879806E+00	113.	0.1663545E-02
Total	0.5220552E+00	126.	0.4143295E-02

Variable	Coefficient	Standard error	t-ratio	Probability x
Constant	5.9612	0.1423	41.883	0.00000
BURGLARY	−0.76784E-03	0.4726E-03	−1.625	0.10700
RAIL	0.41973E-03	0.3140E-03	1.337	0.18396
TAX	0.11853E-03	0.7471E-04	1.586	0.11543
GP	−0.74304E-05	0.3356E-04	−0.221	0.82516
EXAM	−0.10213E-02	0.1021E-02	−1.000	0.31929
POPDEN	0.18625E-05	0.3982E-05	0.468	0.64092
UNEMP	−0.52905E-02	0.2410E-02	−2.195	0.03021
PRECIP	−0.34346E-05	0.2425E-04	−0.142	0.88762
TEMP	0.80578E-02	0.1322E-01	0.610	0.54334
SUN	0.74048E-04	0.6052E-04	1.224	0.22366
LONDON	0.73291E-01	0.1827E-01	4.011	0.00011
WALES	0.86439E-02	0.2008E-01	0.430	0.66766
SCOT	0.37457E-01	0.2648E-01	1.414	0.15998

coefficient in the hedonic wage equation implies that a positive price is being paid in order to obtain incremental units of it. The coefficients on the climate variables in the hedonic wage regression have been set equal to zero because none of the coefficients on climate variables is significantly different from zero in any of the hedonic wage-rate regressions. In principle it would be possible to derive the full implicit prices pertaining to different skill groups by estimating hedonic wage regressions for skilled and manual workers separately. But to do so would require information on average house prices for each group as well.

To begin with, the full implicit price of burglaries is, as expected, negative at -£1.70 per each additional burglary per 1000 people. The coefficient on

Table 1.8 *The Estimated Hedonic Wage Regression:*
Female Non-Manual Workers

Ordinary least squares regression			Dependent variable =		Log WAGE
Observations	=	127	Weights	=	ONE
Mean of left			Standard deviation		
hand side	=	0.6593009E+01	of left hand side	=	0.1336734E+00
Standard deviation			Sum of squares	=	0.4003862E+00
of residuals	=	0.5952513E-01			
R-squared	=	0.8221644E+00	Adjusted R-squared	=	0.8017055E+00
F[13, 113]	=	0.4018603E+02	Probability value		0.0000000E+00
Log-likelihood	=	0.1855239E+03	Restriction (β=0)		
			Log-l	=	0.7586597E+02
Amemiya			Akaike		
probability			information		
criterion	=	−0.2701163E+01	criterion	=	0.3933834E-02

ANOVA source	Variation	Degrees of freedom	Mean square
Regression	0.1851054E+01	13.	0.1423888E+00
Residual	0.4003862E+00	113.	0.3543241E-02
Total	0.2251440E+01	126.	0.1786858E-01

Variable	Coefficient	Standard error	t-ratio	Probability x
Constant	6.4494	0.2077	31.049	0.00000
BURGLARY	−0.44956E-03	0.6897E-03	−0.652	0.51585
RAIL	0.18671E-02	0.4582E-03	4.075	0.00009
TAX	0.20026E-03	0.1090E-03	1.837	0.06891
GP	−0.65332E-04	0.4898E-04	−1.334	0.18489
EXAM	−0.47934E-02	0.1490E-02	−3.217	0.00169
POPDEN	0.12046E-04	0.5812E-05	2.073	0.04049
UNEMP	−0.14870E-01	0.3518E-02	−4.227	0.00005
PRECIP	0.23674E-04	0.3539E-04	0.669	0.50490
TEMP	0.26442E-01	0.1929E-01	1.371	0.17317
SUN	0.11038E-03	0.8832E-04	1.250	0.21397
LONDON	0.10349	0.2667E-01	3.881	0.00018
WALES	0.25895E-01	0.2930E-01	0.884	0.37876
SCOT	0.33449E-01	0.3865E-01	0.865	0.38861

the level of the council tax is, however, unexpectedly signed at +£1.21 and this is almost certainly due to it acting as a proxy for higher spending on local services. The variables indicating the quality of local public goods and services, however, are mixed. Longer GP patient lists have a positive amenity value of +£0.05 per person whereas a negative implicit price would have been more plausible. However, schools with higher exam pass rates are associated with much higher amenity values (+£27.98 per additional one per cent of pupils achieving five or more GCSEs at Grade C or above). The density of railway stations is also regarded as an amenity worth +£3.67 per station per square kilometre multiplied by 100. Similarly, population density is viewed as an amenity worth £0.08 per additional person per square kilometre (presumably there are net benefits from agglomeration which outweigh the disamenities from lower environmental quality).

The negative coefficient on unemployment in the pooled hedonic wage regression is consistent with the literature suggesting that unemployment exerts a downward pressure on wages through competition between workers. But overall, the full implicit price on unemployment is negative, indicating that unemployment is viewed as a disamenity which is consistent with the thesis in Harris and Todaro (1970). Each percentage point increase in the unemployment rate costs £80.59 per household on average.

Turning finally to the climate variables, as anticipated, households in Britain display an aversion to greater precipitation (51 pence per millimetre) and a strong preference for warmer temperatures (£298 per °C) and more sunshine (79 pence per hour). Unfortunately it is difficult to make comparisons between the findings of this study and others with regard to the impact of climate variables on house prices and wage rates. First, the measurement of climate differs between studies and, second, the effect of a marginal increase in the level of a climate variable may differ between countries. All that can be said is that this study, like its predecessors (eg Blomquist et al, 1988, and Maddison and Bigano, 1996), finds a significant role for climate variables in explaining variations in house prices. However, unlike previous studies (eg Hoch and Drake, 1974), there is no role for climate variables in explaining variations in wage rates. In the final chapter of this book these implicit prices are compared to those obtained from the alternative methodology described in Chapter 3 and then used to determine the impact of various climate change scenarios on amenity values in Britain.

Table 1.9 *The Full Implicit Price of Amenities to Households (£:1994 prices)*

Variable	House price	Wage rate	Full price
Burglary rate	−9.79	−8.09	−1.70/burglary/1000 persons
Tax rate	+3.14	+1.93	+1.21/£1 of council tax
GP lists	-0.12	−0.17	+0.05/additional patient/GP
Exam passes	+5.51	−22.47	+27.98/percentage passing five or more GCSEs
Railway stations	+14.31	+10.64	+3.67/station/sq km x 100
Population density	+0.18	+0.10	+0.08/person/sq km
Unemployment	−161.54	−80.95	−80.59/% point
Precipitation	−0.51	–	−0.51/mm
Temperature	+298.28	–	+298.28/°C
Sunshine	+0.79	–	+0.79/hour

CONCLUSIONS

This chapter has demonstrated how households' preferences for climate variables can in theory be deduced from hedonic price regressions for land and labour. Moreover, the analysis offered illustrates that in order to determine the full implicit price of climate amenities it is necessary to consider both markets jointly and that it is even possible that the amenity price gradient may be unexpectedly signed in one of the markets.

The analysis has also discussed the desirability of using data describing residential land prices rather than the price of individual properties with which to uncover the implicit prices of climate variables. The advantage of using data relating to individual properties is that the characteristics of properties (such as physical dimensions, insulation and heating and cooling equipment) themselves change with climate and are otherwise difficult to control for. On the other hand, the existence of building standards may prevent certain characteristics from changing across different sites and census tract data may be more appropriate given that the impact of climate variables on property prices may be small compared to the effect of other non-modelled factors.

In the empirical section of the chapter, the model is estimated on data for Britain and it is shown that the market for housing is segmented along geographic lines. This was presaged by the concerns of theoretical practitioners. The study confirms that when segmentation of the housing market is not dealt with, several important coefficients become statistically insignificant and/or implausibly signed. Similarly, the analysis has also shown that the inclusion of unemployment into the hedonic wage and house price regressions produces results which are consistent both with the view that unemployment leads to wage competition and also that unemployment is viewed as a disamenity by households. There is also evidence that the hedonic price schedule is unstable over different skill groups and that this is because the taste for amenities differs between manual and non-manual workers rather than the underlying assumptions of the hedonic technique being more applicable to one group than the other.

The main deficiency of the study providing this evidence, however, is that relatively few characteristics of workers have been controlled for in the hedonic wage regression. Controlling for more characteristics of the work force and also for differences in the occupational structure is clearly desirable since these are likely to vary considerably across the country. There is also clearly work outstanding in terms of further specifying the levels of other environmental amenities and the quality of local services. This is shown by the unexpected sign of the variable describing the level of local taxes (which was attributed to a failure to control for the levels of all local services). Nevertheless, empirical analysis of census tract data has shown that as a group some climate variables appear to exercise a highly significant and moreover plausible influence over residential property prices in Britain.

NOTES

1   There are no published examples of hedonic analyses involving the price of non-traded goods in the literature. The extent to which the assumption of the 'law of one price' affects the ensuing analysis is an open question.

2   The theory underlying migration studies involves the potential migrant comparing the present value costs and benefits of each location, giving due consideration to costs of relocation (Sjaastad, 1962).

3   It was earlier argued that this is not really a problem since land is seldom sold separately from the structures which stand upon it, but that it would be a serious problem if there were some characteristics of housing which, apart from contributing to the value of a property, systematically varied with amenity levels. Chief among the characteristics of housing is the quantity of land consumed. Plausibly, the quantity of land is a function of the unit price of land, which is in turn a function of the level of environmental amenities. This implies that a failure to control for average floor-space could bias the coefficients on the amenity variables towards zero. Of course this problem can be assumed away by positing that households cannot adjust their consumption of land, but the true relevance of this point can only be assessed through empirical analysis. For an example of a study which uses property prices per unit land area see Maddison and Bigano (1996).

4   These were provided by the Climate Research Unit of the University of East Anglia under the auspices of the TIGER initiative.

5   In other models population density is explicitly endogenous; see for example Steinnes and Fisher (1974). Using the Two Stage Least Squares technique, Nordhaus (1996) treats population density as endogenous regressor in his hedonic wage rate analysis of the amenity value of climate variables. He selects the number of military personnel as an instrument for population density on the grounds that these individuals are not drawn to a given location by virtue of the going wage rate. While the assumption of exogeneity is in principle a testable hypothesis, to do so requires instrumental variables which are not normally available. In any case, the use of instrumental variables would probably frustrate the purposes of using population density as a proxy for other environmental disamenities.

6   Maximizing the expected utility function across the two possible states (employed and unemployed) results in an indirect expected utility function in which the probability of unemployment enters as an exogenous variable. Invoking the equilibrium assumption and totally differentiating the indirect expected utility function with respect to the probability of unemployment, demonstrates that the probability of unemployment must be reflected in regional wage and/or land prices. Dividing this expression throughout by the expected utility of money shows that the marginal willingness to pay for a reduction in the probability of unemployment can be inferred from the hedonic land and wage rate regressions in the usual way.

7   $F_{5,113} = 5.57$; CV at the 99 per cent level of confidence = 3.20.

8   Interestingly Roback finds that when she extends the framework referred to earlier by considering two different types of workers, the hedonic wage gradient of one group of workers depends not only on the productivity effects of climate but also on the preferences of the other group.

9   $F_{40,452} = 5.71$; CV at the 95 per cent level of confidence = 1.39.

10 The exact results are as follows: for non-manual males $F_{3,113} = 1.39$; for non-manual females $F_{3,113} = 3.81$; for manual males $F_{3,113} = 0.40$; and for manual females $F_{3,113} = 2.15$. CV = 2.68 at the 95 per cent level of confidence and CV = 3.81 at the 99 per cent level of confidence.

Chapter 2

THE AMENITY VALUE OF THE CLIMATE: THE HOUSEHOLD PRODUCTION FUNCTION APPROACH

Economic activity is directed at the satisfaction of human wants, the most basic of which is protection from the privations of the climate. But the idea of climate as a direct input to human welfare has received relatively little attention in the climate-change literature. When such issues have been addressed (as in Maddison and Bigano, 1996) it is typically within the context of the hedonic approach. This argues that, if individuals are able to freely select from differentiated localities, then the tendency will be for the benefits associated with them to become collateralized into property prices and wages. In such cases the value of marginal changes can be discerned from the hedonic house and wage price regressions.

But a basic assumption of the hedonic technique is that there are no barriers to mobility which prevent prices changing to reflect the net benefits of a given location. Yet climate variables are undeviating over relatively large distances and the absence of a common language and cultural ties may prevent the net advantages associated with a particular region from being eliminated. These considerations suggest that, except for countries like the US and Italy, alternative methods may be more suitable for estimating the amenity value of climate. This chapter seeks to undertake a systematic examination of the role played by climate in determining consumption patterns using cross-sectional data from 60 different countries and asks under what circumstances it is possible to derive a measure of the amenity value of climate from observed patterns of consumption. The chapter further assumes that these conditions are met and then proceeds to calculate the implicit price of a range of climate variables for each of the 60 countries.

THE HOUSEHOLD PRODUCTION FUNCTION APPROACH

The role of climatic variables in determining patterns of observed expenditures can best be motivated by reference to the Household Production Function (HPF) theory of Becker (1965). In the HPF approach, households combine marketed commodities using a given production technology. These result in a variety of service flows of direct value to the individual concerned. The overall level of utility is maximized by choice of service flows subject to the budget constraint. The price of a service flow is determined by the cost-minimizing combination of marketed commodities necessary to produce a unit service flow.

The majority of work, however, has been on the concepts of 'weak complementarity' and 'weak substitutability' as restrictions on preferences rather than on the production technologies themselves (Maler, 1974, and Feenberg and Mills, 1980). Effectively, both weak complementarity and weak substitutability imply that there must exist either a commodity bundle or a price vector at which the marginal utility afforded by additional amounts of the environmental amenity is zero (this is the 'demand interdependency' assumption of Bradford and Hildebrandt, 1977). Given the assumption of demand interdependency, the economic value of non-marketed environmental amenities can be determined from observing the consumption of marketed commodities.

The intuition underlying demand interdependency is best understood by realizing that integrating the restricted Hicksian demand functions generally results in unknown constants of integration which are a function of the level of the environmental amenity. These constants can only be eliminated if it is known that there is some price vector for which marginal changes in the level of the amenity have no effect upon the expenditure function (see eg Smith, 1991).

There is, however, a class of commonly used utility functions that do not permit the full impact of changes in the levels of environmental amenities to be recovered. These are utility functions in which the environmental amenities form a strongly separable subset. In addition, whether or not demand interdependency holds is itself not a testable hypothesis.

PREVIOUS LITERATURE

Few analyses have attempted to implement the HPF approach to valuing environmental amenities. These analyses have either assumed the existence of an indirect utility function from which demand functions can be derived or, alternatively, have begun by specifying the production technologies and preferences for the various service flows available. The advantage of the latter approach is that it enables the analyst to examine the credibility of the technological relationships being proposed.

That so few researchers should have attempted to implement the HPF methodology is surprising, since others have seen considerable potential in the technique (eg Smith, 1991). Perhaps this is because, whereas researchers have been willing to accept the assumptions underlying techniques such as the hedonic approach, they have been unwilling to make the assumptions associated with the HPF approach (ie the assumptions relating to the restriction on preferences implied by demand interdependency).

Shapiro and Smith (1981) begin with an indirect utility function in which four environmental amenities are included. Expressions for the expenditure shares are derived and the system of equations econometrically estimated using maximum-likelihood methods imposing the cross-equation restrictions suggested by theory. The indirect utility function chosen parodies the transcendental utility function although the variables are not taken in logarithmic form. Using Roy's Identity, the Marshallian demand equations are derived and then the parameters econometrically estimated and inserted into an expression which yields the implicit price of each of the environmental amenities. The results of the exercise are moderately encouraging, especially since only three different commodity bundles are separately identified. This would severely limit the opportunities to identify any relationships between environmental amenities and commodity purchases.

Studies by Math-Tech Inc (1982) and Gilbert (1985), by contrast, both explicitly adopted the HPF approach. Both studies used the Stone-Geary utility function to describe both the aggregated expenditure system as well as the cost sub-functions for each activity. Environmental amenities are brought into the analysis by specifying that the 'subsistence' parameters of the cost sub-functions should be a function of environmental variables. In the context of consumer preferences, the subsistence parameters are often interpreted as a subsistence level of a commodity which must be purchased before any increment to utility is possible. In the context of a production technology, the interpretation is that a change in the provision of one or more of the environmental amenities changes the amount of marketed commodities which must be purchased before any addition to the flow of services is achieved. While both studies consider average expenditure patterns across a number of American cities, unlike the study by Gilbert, the Math-Tech Inc study includes both temperature and rainfall as potential influences on expenditure shares. Neither variable appears to play a significant role.

EXTENDING SYSTEMS OF DEMAND EQUATIONS TO REFLECT THE ROLE OF ENVIRONMENTAL AMENITIES

The procedures used here to incorporate environmental variables into systems of demand equations are identical to the methods used to incorporate demographic variables into systems of demand equations. More specifically, the analysis uses the 'demographic translating' and 'demographic scaling' procedure (see Pollak and Wales, 1981). Using these procedures, the nature of the role played by environmental amenities in combining with marketed goods is made very clear, and established utility functions, whose limitations can constrain the results, can be used to describe the demand for the various service flows.

With 'demographic translating' fixed costs are added to, or deducted from, the operations of the household. Translating replaces the original demand system by:

$$q_i = d_i + q^i \left(p, y - \sum d_k p_k \right)$$

where q is the quantity demanded, p is price, z is a vector of environmental amenities and the d's are the translation parameters given by:

$$d_i = d_i(z) = \sum \eta_i z_i$$

z is a vector of environmental amenities. This corresponds to the direct utility function:

$$u = u(q_1 - d_1, q_2 - d_2, ...)$$

Demographic translating therefore corresponds to a situation in which marketed goods and environmental amenities are combined in a linear production function such that changes in the level of environmental amenities do not alter the price of the service flow, but merely serve to impose a fixed cost upon the household.

In 'demographic scaling' the effective prices of the commodities are increased. In the context of modelling the impacts of household composition the scaling factors can be interpreted in terms of 'adult equivalents'. Scaling replaces the original demand system by:

$$q_i = m_i q^i(p_1, m_1, p_2, m_2, ..., y)$$

where the m's are the scaling parameters given by:

$$m_i = m_i(z) = 1 + \sum \eta_i z_i$$

This corresponds to the direct utility function:

$$u = u(q_1 / m_1, q_2 / m_2, ...)$$

Demographic scaling corresponds to a situation in which a change in the level of an environmental amenity results in a proportionate change in the service flow.

Prior to analysing the data set it is necessary to lend particular functional form to the proposed relationships to provide a basis for estimation. Given the paucity of observations it would be tempting to adopt the Linear Expenditure System (LES) of Stone (1954). The LES is parsimonious in terms of parameters such that a quite detailed commodity disaggregation is possible. This parsimony, however, means that the underlying demand system is not an appealing description of preferences.[1] The Almost Ideal Demand System (AIDS) model of Deaton and Muelbauer (1980) is a more appealing description of preferences, but contains too many parameters. It would be necessary to use a more highly aggregated set of commodities that may prevent the identification of the exact role that environmental amenities play in shaping consumption. As a compromise, the Quadratic Expenditure System (QES) of Howe et al (1979) is employed. The chief characteristic of this system of the demand equations is that it allows the Engel curves to be a quadratic function of income levels. The QES system also includes the LES as a special case.[2]

Table 2.1 *Definition of Commodity Aggregates for the QES Model*

Commodity aggregate	Commodities included
Food	Bread, rice and cereals
	Meat, fish, dairy products and fats
	Fruit and vegetables
	Coffee, tea, cocoa and sugar
	Soft drinks
Clothing	Clothing
	Footwear
Shelter	Rents and imputed rents
	Repair and maintenance of housing
	Electricity
	Gas, oil and other fuels
	Household goods and appliances
	Furniture, fixtures and household textiles
	Glassware and cutlery
	Cleaning and domestic services
Other	Alcoholic drinks
	Tobacco
	Personal transport equipment
	Gasoline, motor oils, grease and other running costs
	Fares
	Telephone and postal services
	Radio and television
	Sports equipment
	Books, newspapers and magazines
	Education
	Pharmaceutical products, medical goods and services
	Toiletry articles, beauty and hairdressing
	Jewellery
	Restaurants and hotels
	Financial services

Table 2.2 *Records Used to Compute Climate Averages for Different Countries*

Country	City or record	Population ('000)
US	New York	18,120
	Los Angeles	13,770
	Chicago	8181
	San Francisco	6042
	Philadelphia	5963
	Detroit	4620
	Dallas	3766
	Boston	3736
	Washington DC	3734
	Houston	3642
	Miami	3001

Table 2.2 *(Continued)*

Country	City or record	Population ('000)
	Cleveland	2769
	Atlanta	2737
	Saint Louis	2467
	Seattle	2421
	Minneapolis	2388
	San Diego	2370
	Baltimore	2343
	Pittsburgh	2284
	Phoenix	2030
Belgium	Brussels	
Denmark	Copenhagen	
France	Paris	8510
	Lyon	1170
	Marseilles	1080
Germany	Berlin	3301
	Hamburg	1594
	Munich	1189
Greece	Athens	
Ireland	Dublin	
Italy	Rome	2817
	Milan	1464
	Naples	1203
Luxembourg	Luxembourg	
Netherlands	De Bilt	
UK	London	
Austria	Vienna	
Finland	Helsinki	
Hungary	Budapest	
Norway	Oslo	
Poland	Warsaw	
Portugal	Lisbon	1612
	Oporto	1315
Spain	Madrid	3123
	Barcelona	1694
Yugoslavia	Belgrade	
Botswana	Francistown	

Table 2.2 *(Continued)*

Country	City or record	Population ('000)
Cameroon	Douala	
Ethiopia	Addis Ababa	
Ivory Coast	Abidjan	
Kenya	Nairobi	
Madagascar	Antananarivo	
Malawi	Lilongue	
Mali	Bamako	
Morocco	Rabat	
Nigeria	Lagos	1097
	Ibadan	1060
Senegal	Dakar	
Tanzania	Dar es Salaam	
Tunisia	Tunis	
Zambia	Lusaka	
Zimbabwe	Harare	
Israel	Haifa	
Hong Kong	Hong Kong	
India	Calcutta	9194
	Bombay	8243
	Delhi	5729
	Madras	4289
	Bangalore	2922
	Ahmadabad	2548
	Hyderabad	2546
Indonesia	Jakarta	7348
	Surabaya	2224
	Medan	1806
Japan	Tokyo	11,829
	Yokohama	2993
	Osaka	2636
	Nagoya	2116
People's Republic of Korea	Pyongyang	
Pakistan	Karachi	5208
	Lahore	2953
	Hyderabad	1104
Philippines	Manila	

Table 2.2 *(Continued)*

Country	City or record	Population ('000)
Sri Lanka	Colombo	
Argentina	Buenos Aires	10,728
	Cordoba	1055
	Rosario	1016
Bolivia	La Paz	
Brazil	Rio de Janeiro	11,141
	Belo Horizonte	3446
	Recife	2945
	Porto Alegre	2924
	Salvador	2362
Chile	Santiago	
Colombia	Bogota	4185
	Medellin	1506
Costa Rica	San Jose	
Dominica	Rouseau	
Ecuador	Guayaquil	1301
	Quito	1110
El Salvador	San Salvador	
Guatemala	Guatemala City	
Honduras	Tegucigalpa	
Panama	Balboa Heights	
Paraguay	Asuncion	
Peru	Lima-Callao	
Uruguay	Montevideo	
Venezuela	Caracas	3247
	Maracaibo	1295
Canada	Toronto	3427
	Montreal	2921
	Vancouver	1381

Source The Phillips Atlas and Landsberg (1969)

The expenditure share equations associated with the QES are:

$$\frac{p_i q_i}{y} = \frac{p_i \gamma_i}{y} + \beta_i \left(1 - \frac{\sum p_i \gamma_i}{y}\right) + \left(\delta_i \frac{p_i}{y} - \beta_i \frac{\sum p_i \delta_i}{y}\right) \Pi \left(\frac{p_i}{y}\right)^{-2\beta_i} \left(1 - \frac{\sum p_i \gamma_i}{y}\right)^2$$

with the adding-up restriction that:

$$\sum \beta_i = 1$$

DATA SOURCES

The price and expenditure data are taken from the 1980 International Comparisons Project (ICP) involving 60 different countries (Kravis et al, 1982). This project provides quinquennial information on per capita consumption patterns in national currencies and Purchasing Power Parities (PPPs) in terms of national currencies per US dollar for 108 consistently defined final consumption commodities.[3] For the purposes of this study a four-commodity disaggregation is adopted: food and drink; clothing and footwear; housing, household goods and services and fuel (shelter); and other goods and services. The components of these commodity aggregates are given in Table 2.1. Prices are aggregated using quantity weights (national expenditures divided by PPPs).[4 and 5]

Climate records for each country's major cities are taken from Landsberg (1969). To arrive at a single index for each country, the records for individual cities are population-weighted.[6] The records used to determine the indices for each country are shown in Table 2.2. Two climatic variables are included in the analysis: the annually averaged mean temperature (denoted by T) expressed in terms of degrees centigrade, and annual average precipitation (denoted by R) measured in terms of millimetres. Substantial variation in annual average temperatures and rainfall is observed over the data set. The scaling function for commodity i is given by:

$$m_i = 1 + \eta_{1i}T + \eta_{2i}T^2 + \varphi_{1i}R + \varphi_{2i}R^2$$

while the translating function for commodity i is given by:

$$d_i = \eta_{1i}T + \eta_{2i}T^2 + \varphi_{1i}R + \varphi_{2i}R^2$$

ESTIMATION AND RESULTS

In order to overcome likely problems associated with heteroscedasticity, the system of n-1 equations is estimated in share form by a maximum-likelihood technique.[7] For the sake of comparison, each of the two alternative ways of introducing environmental variables into the QES model was estimated alongside the straight LES and QES models. The corresponding log-likelihoods are presented in Table 2.3. It is immediately apparent that compared to the QES model, the LES model offers a significantly worse fit to the data.[8]

The main point of interest, however, is to establish whether the exclusion of all climate variables from the QES model represents a statistically significant restriction. Using a Likelihood Ratio test, the null hypothesis of no role for the climatic variables can be rejected against both the translating procedure and

Table 2.3 *Summary Statistics for Four Alternative Models of Demand*

Model	Log likelihood	Number of parameters
LES	309.453	7
QES	334.310	11
QES with translating procedure	360.997	27
QES with scaling procedure	357.853	27

Table 2.4 *Parameter Estimates of the QES Model of Consumer Demand Employing the Quadratic Translating Procedure*

Commodity	Parameter	Estimate	t-statistic
Food	γ	−141.541	(−0.749)
	β	0.454	(20.653)
	δ	−0.505E-4	(−3.100)
	η_1	−18.367	(−0.935)
	η_2	0.860	(1.710)
	φ_1	0.194	(2.006)
	φ_2	−0.720E-4	(2.147)
Clothing	γ	−95.411	(−2.127)
	β	−	−
	δ	−0.322E-5	(−1.317)
	η_1	−12.681	(−2.713)
	η_2	0.357	(2.971)
	φ_1	0.019	(0.539)
	φ_2	−0.951E-5	(−0.693)
Shelter	γ	3.497	(1.443)
	β	0.174	(13.942)
	δ	0.289E-5	(0.460)
	η_1	−10.138	(−2.003)
	η_2	0.321	(2.049)
	φ_1	0.099	(1.225)
	φ_2	−0.396E-4	(−1.176)
Other	γ	−1.140	(−0.234)
	β	0.292	(16.123)
	δ	0.302E-4	(2.682)
	η_1	−29.582	(−2.183)
	η_2	0.876	(2.836)
	φ_1	0.270	(1.618)
	φ_2	−0.845E-4	(−1.319)

the scaling procedures at the 99 per cent level of confidence.[9] The translating model, however, has a somewhat higher log-likelihood function than its scaling counterpart and accordingly becomes the focus of attention for the remainder of the chapter. With the translating procedure, climate variables represent fixed costs to the individual, so that any deterioration or improvement in the climate is likely to have a highly regressive or progressive impact.

Eliminating either the temperature variables or the precipitation variables from the QES translating procedure results in a loss of fit significant at the 99 per cent level of confidence. Eliminating the quadratic variables results in an equally significant loss of fit implying that the marginal willingness to pay for climate variables depends upon the current climate.[10] Finally, imposing those restrictions necessary to reduce the QES model to the LES model with translating is also rejected.[11]

The parameter estimates of the QES model with climate variables as quadratic translating variables are given in Table 2.4. Turning to the individual share equations, it appears to be difficult to explain the cross-country variation in the purchase of some commodities unless account is taken of differences in climate (see Table 2.5). This is particularly true of expenditure on clothing and shelter where the greatest proportionate increases in fit are observed when climate variables are added.

Table 2.5 *The Relative Explanatory Power of the Constrained and Unconstrained (Translating Procedure) QES Model of Consumer Demand*

Commodity group	R^2 statistic for the QES model	R^2 statistic for the translating procedure
Food	0.736	0.837
Clothing	0.004	0.227
Shelter	0.239	0.432
Other	0.710	0.755

DISCUSSION

Table 2.6 records the per capita willingness to pay (corresponding to the reduction in fixed costs) associated with a 1°C temperature increase and a 1 mm increase in precipitation. These values are expressed in terms of 1980 US dollars converted by means of PPP exchange rates for each of 60 different countries.

Most of the countries in Europe and North America appear to benefit substantially from an increase in temperature. Canada, Norway and Finland in particular appear willing to pay sizeable amounts for a 1°C increase in annually averaged temperature. The US, Italy, Portugal and Spain, on the other hand, appear to be close to climatically optimal conditions. Only Greece is significantly affected among the European countries, but her losses are small in comparison with the gains enjoyed elsewhere. All the countries in Europe and North America exhibit a positive willingness to pay to avoid any increase in precipitation, with the figure reaching 43 cents per millimetre in the case of Denmark.

Turning to the Indian subcontinent, a completely different picture emerges. Pakistan, India and Sri Lanka all lose from a 1°C rise in annually averaged mean temperature. Although the amounts involved are small in absolute terms, they are significant insofar as these countries are so populous and the current levels of income so low. Except for the People's Republic of Korea, the

Table 2.6 *The Estimated Per Capita Willingness to Pay for Climate Variables (1980 US$)*

Country	Willingness to pay for a 1°C rise in annually averaged mean temperature	Willingness to pay for a 1 mm increase in precipitation
US	3.59	−0.21
Belgium	27.40	−0.27
Denmark	39.26	−0.43
France	17.07	−0.38
West Germany	36.16	−0.38
Greece	−15.24	−0.36
Ireland	25.90	−0.28
Italy	−3.43	−0.19
Luxembourg	32.62	−0.32
Netherlands	29.25	−0.31
UK	21.38	−0.37
Austria	26.56	−0.34
Finland	61.14	−0.38
Hungary	6.90	−0.13
Norway	58.82	−0.38
Poland	21.84	−0.24
Portugal	−3.11	−0.12
Spain	−0.14	−0.35
Yugoslavia	8.86	−0.22
Botswana	−23.89	−0.30
Cameroon	−55.40	1.08
Ethiopia	−3.19	−0.03
Ivory Coast	−57.38	0.21
Kenya	−10.86	−0.12
Madagascar	−13.02	−0.02
Malawi	−15.08	−0.17
Mali	−35.93	−0.07
Morocco	−10.72	−0.25
Nigeria	−75.73	0.05
Senegal	−32.90	−0.24
Tanzania	−55.48	−0.15
Tunisia	−9.07	−0.25
Zambia	−32.92	−0.29
Zimbabwe	−14.45	−0.18
Israel	−28.25	−0.25
Hong Kong	−26.66	0.22
India	−27.99	0.08
Indonesia	−24.37	−0.07
Japan	−3.34	0.08
People's Republic of Korea	15.72	−0.13
Pakistan	−17.84	−0.14
Philippines	−25.05	0.11
Sri Lanka	−17.09	0.11

Table 2.6 *(Continued)*

Country	Willingness to pay for a 1°C rise in annually averaged mean temperature	Willingness to pay for a 1 mm increase in precipitation
Argentina	−12.82	−0.24
Bolivia	13.06	−0.20
Brazil	−20.52	−0.01
Chile	−1.96	−0.32
Colombia	−4.92	−0.05
Costa Rica	−22.27	0.11
Dominica	−35.96	0.14
Ecuador	−17.45	−0.08
El Salvador	−28.22	0.08
Guatemala	−15.54	−0.02
Honduras	−21.93	−0.12
Panama	−39.39	0.10
Paraguay	−30.35	−0.02
Peru	−12.87	−0.26
Uruguay	−6.19	−0.15
Venezuela	−39.42	−0.22
Canada	32.68	−0.15

remainder of Asia also loses from a 1°C rise in temperature although the per capita loss experienced by Japan is very small.

All the African countries in the data set appear to lose from temperature increases. Cameroon, Ivory Coast, Tanzania and, in particular, Nigeria are worst affected. Increases in annually averaged mean temperatures of 1°C would cost in excess of US$50 per capita and in the case of Nigeria more than US$75. Mali, the hottest country in the entire data set, exhibits a willingness to pay to avoid a 1°C rise in temperature of US$36. Paradoxically Ethiopia, one of the poorest countries in Africa, is the least affected due to the relatively cooler climate of that country. As with Asia, there are examples of countries with very high precipitation levels exhibiting an appetite for yet more rainfall. An extreme case is Cameroon, whose annual rainfall exceeds 4000 millimetres.

Turning to Central and South America, Argentina, Chile, Colombia, Peru and Uruguay are not much affected by a 1°C rise in temperature due to the elevation of those countries or their higher latitudes. Such is the elevation of Bolivia, there is an actual gain from a 1°C rise in temperature. Closer to the equator, however, countries such as Panama and Venezuela are adversely affected, with a negative willingness to pay of US$39 per capita. Peru, the driest country in the data set, exhibits a marginal willingness to pay to avoid increases in precipitation of 26 cents per millimetre.

The change in the cost of living caused by increases in temperature results in a loss of welfare for poor countries whilst the rich countries gain. Considering the concept of an 'optimal' climate, it is necessary to understand that such a climate depends upon relative prices. However, there are no examples of countries with an annually averaged mean temperature in excess of 15°C being willing to pay anything for further increases. The picture for precipitation is more complex. At low levels of precipitation, precipitation is viewed as a disamenity. The marginal price of precipitation, however, diminishes as precipitation increases and at very high levels of precipitation it becomes an amenity. This is hard to explain in terms other than the limitations of a functional form that compels marginal willingness to pay to be a linear function of current levels of precipitation. Perhaps the reality is that marginal willingness to pay for precipitation approaches zero asymptotically as precipitation increases.

CONCLUSIONS

Differences in the cost of living between urban areas and rural areas and differences in the distribution of incomes mean that it will always be difficult to explain differences in average per capita patterns of consumption between countries. Furthermore, the climate variables used in this analysis have been averaged over large and diverse geographical areas.

Given the limitations of the data used in this empirical analysis it is perhaps all the more surprising that one is able to produce a set of results so much in line with expectations. Climate represents a highly significant determinant of expenditure patterns. Temperatures in North America and northern Europe are too cold, but the temperatures that characterize the Mediterranean are just right. Precipitation is a disamenity although one that diminishes in value as precipitation levels increase. These results imply that it would be worthwhile to collect the requisite micro-economic data from narrowly defined climatic regions.

Were such data ever to become available, it would be of interest to observe the extent to which differing specifications of the underlying demand system and different commodity aggregations effect the implicit prices for the environmental amenities. It would also be interesting to examine the use of other climate variables representing perhaps the variation of the climate across the seasons. Finally, and perhaps most importantly, because significant climate change is not expected until midway through *this next century*, some knowledge of the role of income in determining the amenity value of climate is required. In the current analysis it seems that climate imposes a fixed cost on society and therefore its significance is bound to diminish. Rich people, it seems, do not care for the climate.

In the meantime, these preliminary results suggest that global temperature increases are likely to confer benefits on cold northern countries. In hot tropical countries, by contrast, any increase in temperature is likely to reduce welfare and result in a large change in the cost of living. Increased precipitation is more likely to impose a cost on countries in the north.

NOTES

1   Brenton (nd) for example finds the fit afforded by the LES to international consumption data to be especially poor and is compelled to split his data set into 'rich' and 'poor' countries.

2   For an attempt to incorporate demographic variables into the QES system see Pollak and Wales (1980). For an attempt to estimate the QES on international consumption data see Pollak and Wales (1987).

3   Countries measure quantities by index numbers defined in value units based on the country's national currency. Pooling consumption data from different countries requires transforming them to common units using different transformations for each good. The required transformation factors are called Purchasing Power Parities. Except under the assumption of perfect goods arbitrage, these cannot be inferred from market exchange rates.

4   The ICP consumption data for 1980 have already been analysed by Brenton (nd) who finds that the degree of fit afforded by a LES system is extremely disappointing. He estimates the LES model, again dividing the sample into rich and poor countries. The hypothesis of parameter homogeneity between the two sets of countries is overwhelmingly rejected. Brenton conducts a similar exercise using the AIDS model and finds that is also provides a poor fit to the data so that the hypothesis of parameter homogeneity between rich and poor nations can again be rejected.

5   Apart from the literature on attempting to derive implicit prices for environmental amenities using the HPF approach, there is also literature on attempts to explain variations in cross-country patterns of consumption: this is briefly surveyed in Selvanathan and Selvanathan (1993). The attraction lies in the fact that there is a large variation in both incomes and prices relative to time series studies or cross-sectional studies undertaken within a single country. This literature was in part stimulated by the controversial hypothesis of Stigler and Becker (1977) that tastes are the same across different countries. Pollak and Wales (1987) and Selvanathan and Selvanathan (op cit) test the hypothesis that tastes are identical across countries by pooling international data and testing the acceptability of restricting the equations of different countries to share common parameters. In both cases the hypothesis of common tastes is rejected. Of course, the results of such tests do not determine whether tastes differ between countries or whether consumption patterns differ as a consequence of differences in household production technologies or the endowment of environmental amenities used in household production processes. In the present work, common tastes is a maintained hypothesis.

6   Even this procedure may not work well in a country like India where the majority of the population lives in rural areas.

7   The adding-up property implies singularity of the variance-covariance matrix. This can be dealt with by dropping one of the budget share equations. The estimation technique is such that the parameter estimates that emerge are not affected by the choice of which equation to drop. The value of the last b parameter is determined from the adding-up constraint.

8   The implied restrictions have a χ^2 statistic of 49.714 against a critical value of 13.277 with four degrees of freedom.

9   The implied χ^2 statistics are 53.374 and 47.086 respectively against a critical value of 32.000 with 16 degrees of freedom.

10 The implied χ^2 statistics are 28.522, 30.314 and 26.384 respectively against a critical value of 21.666 with eight degrees of freedom.

11 The χ^2 statistic is 59.708 against a critical value of 13.277 with four degrees of freedom.

Chapter 3

THE IMPACT OF CLIMATE CHANGE ON AGRICULTURE IN BRITAIN

It is widely believed that climate change could have a significant impact upon agricultural productivity, both positive and negative, and much research effort has been spent in constructing models of the agricultural sector to investigate various climate change scenarios (Parry, 1990). The most rudimentary approach is an empirical representation of the climate-yield relationship. This technique uses multiple regression analysis to explain some aspect of production (normally yield) by reference to fertilizer applications, a time trend for technology and a set of weather variables. Using the estimated model it is possible to simulate the effect of changed weather patterns on yield. Explicitly, however, these analyses provide little information on the cost to agriculture of climate change. In particular, to regard them as indicative of the likely impact of climate change on agricultural productivity would be to adhere to the 'dumb farmer hypothesis' in which farmers fail entirely to respond to or even apparently notice the change in climate. One must at least have regard for the likely alterations to agro-practices that could be expected in response to a permanent change in climate. Permanent changes in climate will also cause agriculture to migrate to more suitable areas and a more sophisticated model of land allocation is required to address such considerations.

In land allocation models producers respond to changed climates by a changed overall pattern of production as well as regional changes in production. An example of such a model for England and Wales is found in Hossell et al (1993). The basic land allocation model is a linear programme which allocates land between a choice set of activities in an attempt to maximize economic surplus. The constraints relate to the contents of the choice set, the availability of suitable land and the price of factor inputs. Using such models, investigators have attempted to predict the efficient response of farmers in particular regions, frequently taking equilibrium carbon dioxide doubling as a reference scenario. The technique is to interpret climate change as an alteration in the availability of land suitable for particular activities and then measure the difference in realizable economic surplus in the face of climate change.

This modelling strategy has been challenged by Mendelsohn et al (1993, 1994), who argue that such models typically fail to find the highest value use for land by considering only a limited number of alternative uses and that they inevitably neglect a whole range of adaptations and changes in farming practices. They suggest that given the number of changes which might occur, the land allocation approach cannot as a practical matter account for them all. Instead, the authors propose a 'Ricardian' (or hedonic) approach to measuring the potential impact of climate change on US agriculture which seeks current day analogues for such change. They argue that the value of land derives from the sum of discounted profits which may be derived from its use. Anything which affects the productivity of land will be reflected in the purchase price. In principle, therefore, land prices contain information concerning the value of climate as a characteristic of the land. Rather than studying the yield changes of specific crops, the Ricardian approach examines how the climate in different locations is consolidated into different rents for land. The approach also accounts for the impact of climate on pests and diseases as well as changes in practice and technique.

The work of Mendelsohn et al confirms that climate variables exert a significant influence on land prices along with other characteristics of the land such as the quality of the soil, its elevation and drainage. The authors argue that the adaptations possible are such that the impact of climate change on the US agricultural sector would be negligible or even mildly beneficial. Recently, Dinar et al (1998) and Evenson and Alves (1998) have successfully used the hedonic technique to measure the amenity value of climate variables to agriculture in India and Brazil respectively. The main limitations of using the hedonic technique to determine the impacts of climate change on agriculture are, first, that the approach cannot value changes due to elevated CO_2 concentrations[1] and, second, that it does not allow for price effects arising from changed levels of inputs and outputs.

The hedonic technique has also been employed by Palmquist and Danielson (1989) to measure the value of erosion control and drainage, and by Miranowski and Hammes (1984) to determine the implicit prices for soil characteristics in Iowa.

This chapter attempts to replicate the findings of the preceding Ricardian studies in the European context. What distinguishes this research is that it deals with the price of individual plots of agricultural land rather than county-wide averages. This makes it possible to test a number of additional hypotheses which, apart from being interesting in their own right, might lend additional credence to the technique. It also means that the climate of individual plots of land can be determined more precisely rather than computing county-wide averages for each climate variable.

THE HEDONIC TECHNIQUE

In the model developed below, agricultural production is assumed to be a constant returns to scale activity and agricultural production a perfectly tradeable commodity such that a unit cost function can be defined for agricultural production. This price must equal the world price in all locations, for if it did not the price of agricultural land and the wage rate for agricultural workers would adjust under pressure from farmers vacating high-cost production sites in favour of low-cost ones. Underpinning this is the assumption that farmers are perfectly informed about the characteristics of different tracts of land and the assumption that they are actually able to move.[2] The unit cost function (A) is:

$$A(w, r^a, z) = 1$$

where w is the wage rate, r^a is the price of agricultural land and z is a characteristic of the land (such as the climate). The unit price of agricultural output is normalized to unity. Totally differentiating the unit cost function gives:

$$\frac{\partial A}{\partial w}\frac{dw}{dz} + \frac{\partial A}{\partial r^a}\frac{dr^a}{dz} + \frac{\partial A}{\partial z} = 0$$

Rearranging the equation and substituting expressions obtained from differentiating the unit cost identity gives:

$$-A_z = N^a \frac{dw}{dz} + L^a \frac{dr^a}{dz}$$

where N^a is the number of agricultural workers and L^a is the amount of agricultural land available. Hence the marginal change in production costs with respect to the climate amenity is, with a change in sign, equal to the marginal change in labour costs multiplied by the current labour force plus the marginal change in the cost of the land multiplied by the current area of land.

Clearly it does not suffice to estimate only the hedonic land price schedule and differentiate it with respect to the characteristics to obtain the implicit price of the amenities. This captures only a subset of the value of the climate variables to agriculture. One has also to account for the fact that the price of labour in general depends upon climate amenities too. However, despite the potential importance of accounting for climate induced compensating wage differentials for agricultural workers, the subsequent section deals exclusively with the estimation of the hedonic agricultural land price schedule, and then goes on to derive the implicit prices of climate variables from this schedule. The reason for the focus on agricultural land prices is that detailed information concerning geographical differences in the wages of workers in the agricultural sector is not available for England and Wales. For this reason alone the exercise cannot be regarded as producing definitive estimates of the implicit prices of climate variables.[3] Of course, this does not preclude any future analysis from looking at wage rates.

Although the question of what constitutes the appropriate functional form of the hedonic land price schedule is typically viewed as being solely an empirical question, Parsons (1990) shows that there are theoretical reasons for supposing that the price of locational characteristics such as climate increases proportionately with farm size; that the hedonic price function should be additively separable in terms of the structural attributes of the land; and that the price per acre should be independent of the quantity of land. The logic of these conditions is obvious. For example, if the price per acre were found to depend on the number of acres, or the price of locational characteristics was not proportional to plot size, then it would mean that profits could be earned from either dividing or combining plots of land. Likewise, the price function should be additively separable in structural characteristics since it is always possible to buy an undeveloped plot and build on it at a cost which is usually independent of either the size of the plot or its locational characteristics (see Appendix 1 for details).

The assumptions underpinning these theoretical restrictions are nothing more than the assumption of perfect competition in the market for land and attributes as well as zero transaction costs (so that the number of sales does not itself enter as an argument in the seller's profit function). If, however, transaction and bargaining costs form a significant cost then land cannot be costlessly repackaged and otherwise identical plots of different sizes may well be sold for different unit costs. There do not appear to be any tests either of repackaging or additive separability in the hedonic literature so one is offered below using farmland price data.

THE DATA SET

The empirical exercise draws on data on land transactions in England and Wales taken from *Farmland Market* (1994). This journal contains a county-by-county record of almost 500 transactions in farmland during the first six months of 1994.[4] The journal records the location, acreage, whether the property sold has vacant possession, along with other important details concerning dwellings and other buildings included in the sale. The prices shown are in some cases the results of sales auctions, although many are sales by private treaty. In the latter case the guide (brochure) price is quoted along with some indication of whether the actual sale price was above, below or close to the guide price. The Ordnance Survey (1992) *Gazetteer of Great Britain* is used to determine the grid reference location of the individual properties from the given address, either to a named farm or the nearest named settlement.

A set of land quality variables are included in the data set by indirectly using the five kilometre grid square *Agricultural Land Classification Of England and Wales* (Ministry of Agriculture, Fisheries and Food, 1988). This classifies land into one of seven grades according to the extent to which its physical characteristics impose long-term limitations on agricultural use. This grading system does not, it is claimed, necessarily reflect the current economic value of the land although it is almost invariably taken that Grade 1 land is the best since there are few limitations to its use. Grade 7 is urban land, non-agricultural land or, as in this case, land which was not surveyed. The principal physical factors included in the grading system are those relating to the site (gradient, microrelief and flood risk), soil (texture, structure, depth and stoniness) and climate. That climate variables should have been incorporated in this classification is a nuisance since it would appear to leave the land classification correlated with the individual climate variables discussed below. Simple correlations between the climate variables and the soil grades, however, suggest that this is unlikely to be a cause for concern. Although the soil

Table 3.1 *Definition of Variables Contained in the Data Set*

Variable	Definition
PRICE/ACRES	Price per acre (£)
ACRES	Number of acres
POSSESS	Dummy variable which takes the value unity if the property is vacant possession; zero otherwise
PRIVATE	Dummy variable which takes the value unity if the property was sold by private treaty; zero otherwise
BEDROOMS	Number of bedrooms in dwellings of a specified size
MILK	Number of milk production quotas offered with the property (×1000 litres)
COTTAGES	Number of dwellings of an unspecified size
SOIL1	Dummy variable which takes the value unity if land is officially classified as Grade 1; zero otherwise
SOIL2	Dummy variable which takes the value unity if land is officially classified as Grade 2; zero otherwise
SOIL3	Dummy variable which takes the value unity if land is officially classified as Grade 3; zero otherwise
SOIL4	Dummy variable which takes the value unity if land is officially classified as Grade 4; zero otherwise
SOIL5	Dummy variable which takes the value unity if land is officially classified as Grade 5; zero otherwise
SOIL6	Dummy variable which takes the value unity if land is officially classified as Grade 6; zero otherwise
FD	30-year average number of frost days
TEMP	30-year average temperature (°C)
WIND	30-year average wind speed (m/s)
PRECIP	30-year average precipitation (mm)
SUN	30-year average hours of sunshine
REH	30-year average relative humidity (percentage)

quality data have the form of a single variable taking the integer values 1–7, the information is incorporated into the ensuing statistical analysis by means of six dummy variables. The reason is that the increase in value of the dependent variable to a change in the grading of soil quality need not be a constant.

Climate variables are calculated on a 10 kilometre grid square basis.[5] Observations can thus be fitted into one of several thousand different climatic regions. Various summertime (April to September) and wintertime (October to March) 30-year climate averages (1961–1990) are matched to the location of the farmland along with the land classification. It is difficult to exclude any climate variable on *a priori* grounds and all those variables contained in the climate database are included as explanatory variables in the statistical analysis which follows. Unfortunately, soil moisture deficits, which are known to be important for crop yields, are not included in the database. The information currently contained in the data set is described in Tables 3.1 and 3.2.

Table 3.2 *Characteristics of the Data Set*

Variable	Mean	Standard deviation	Minimum	Maximum
PRICE/ACRES	2641.5525	2345.90942	176.05634	32545.45508
ACRES	130.07407	207.35947	7.00000	2090.00000
POSSESS	0.98025	0.13014	0.00000	1.00000
PRIVATE	0.62963	0.48222	0.00000	1.00000
BEDROOMS	2.27407	3.77937	0.00000	43.00000
MILK	15.78849	84.41078	0.00000	738.44000
COTTAGES	0.25432	1.33573	0.00000	16.00000
SOIL1	0.022222	0.14759	0.00000	1.00000
SOIL2	0.13086	0.33767	0.00000	1.00000
SOIL3	0.44938	0.49805	0.00000	1.00000
SOIL4	0.18025	0.38487	0.00000	1.00000
SOIL5	0.029630	0.16977	0.00000	1.00000
SOIL6	0.098765	0.29872	0.00000	1.00000
FD (Summer)	19.66116	4.18204	7.09800	29.89800
FD (Winter)	82.90919	13.56708	42.90000	111.19800
TEMP(Summer)	12.71263	0.67990	10.06700	13.90000
TEMP (Winter)	5.70588	0.72960	3.16700	8.08300
WIND (Summer)	4.39904	0.28339	3.91700	5.40000
WIND (Winter)	5.19988	0.57672	4.21700	7.15000
PRECIP (Summer)	387.94078	91.64752	274.09799	754.30200
PRECIP (Winter)	495.88036	177.53136	262.69800	1143.79797
SUN (Summer)	1031.6150	76.14391	861.90002	1224.00000
SUN (Winter)	431.64661	34.60698	342.49802	517.69800
REH (Summer)	82.31932	2.09038	77.51700	87.30000
REH (Winter)	89.62510	1.56296	85.21700	94.06700
Number of observations	405			

The relevance and expected impact on land values of some of the recorded variables may need further explanation. First, many of the farms or land were sold together with other assets of worth. Virtually all farmsteads have large farmhouses attached to them and often labourers' cottages or holiday homes too. By contrast, farm buildings are not consistently recorded in the source and are consequently excluded from the data set, which represents an unfortunate closure on the information available. Milk quotas permit a farmer to sell up to a given volume of milk and any excess production must be discarded. These milk quotas are thus of considerable value to dairy farmers and are sometimes sold along with the farm. Farm equipment and livestock are, by contrast, usually sold separately. In England and Wales only a very few farms are sold as tenanted property: it seems likely that the many regulations governing such property reduce its value in the eyes of an institutional investor.[6] The data set thus records whether a property is sold as a regulated tenancy or not.

EMPIRICAL RESULTS

Before considering the task of measuring the impact of climate on agricultural land prices, it is necessary to check whether the theoretical restrictions suggested by Parsons (1990) hold in the context to the current data set. Consider the case of the simple hedonic price equation below from which the variables describing the irreproducible characteristics of the land such as climate have, for the sake of brevity, been omitted. The theoretical restrictions suggest[7] that the coefficients on the terms relating to the size of the plot (ie β_1 and β_2) as well as the coefficients on those structural attributes not divided through by the number of acres (ie β_3, β_4 and β_5) could be set equal to zero without significant loss of fit:

$$Price_i \ / \ Acres_i = \alpha + \beta_1 Acres_i + \beta_2 Acres_i^2 + \beta_3 Cottages_i +$$
$$\beta_4 Bedrooms_i + \beta_5 Milk_i + \beta_6 Cottages_i \ / \ Acres_i +$$
$$\beta_7 Bedrooms_i \ / \ Acres_i + \beta_8 Milk_i \ / \ Acres_i + \ ...$$

In fact estimating the model with Ordinary Least Squares shows that β_1 is negative and significant and β_2 is positive and significant (see Table 3.3) suggesting that the price per acre first falls and then rises with plot size.[8] This implies that there are profits to be made from repackaging parcels of land. Furthermore, one of the structural attributes is a significant determinant of price per acre. More specifically it appears that milk production quotas, even though they may be sold separately on the open market, are somehow more valuable when they are attached to smaller parcels of land.

These findings suggest that substantial transaction costs make it difficult to put much weight on arguments based on arbitrage. Alternatively the failure of Parsons' hypotheses might be a reflection of the functional form used. For example, one of the implications of the purely linear model is that the marginal value of any one of the characteristics is independent of the level of any other characteristic, implying for example that the value of climate amenities is independent of the underlying quality of the land. It seems preferable that the remaining hypotheses should be tested and amenity values inferred from a model whose functional form is based solely on a goodness-of-fit criterion rather than theoretical considerations.

The only concession made to functional form was to attempt the log transformation of the dependent variable.[9] In the semi-log model the marginal value of any one characteristic depends on the levels of all other characteristics, which is contrary to the theoretical restrictions suggested by Parsons. Using the method described by Maddala (1977) it was found that even though the R-squared statistic is much lower, the semi-log model was more likely to have generated the observed data. The semi-log model succeeds in explaining 60 per cent of the variation in the data (see Table 3.3). Furthermore, restricting the model by excluding all the climate variables yields a χ^2 test statistic of 63.02 against a critical value of 26.22 at the 1 per cent level of confidence with 12 degrees of freedom. Hence the main thesis of the existing research is upheld here too: the amenity values of climate variables are to some extent embedded in farmland prices.

Table 3.3 *The Hedonic Price Equation*

Dependent variable	Price/acre	log (Price/acre)
C	9021.97	15.1824
	(1.19925)	(4.36678)
ACRES	−2.84514	−0.957948E-03
	(−4.08103)	(−4.07162)
ACRES^2	0.131640E-02	0.536776E-06
	(4.34052)	(3.97679)
COTTAGES	247.932	0.018319
	(1.50703)	(2.56627)
BEDROOMS	44.9354	0.567102E-03
	(1.13483)	(3.00638)
MILK	−3.33883	0.068849
	(−4.29750)	(2.21697)
POSSESS	1094.93	9.51889
	(4.76009)	(10.8072)
PRIVATE	−124.636	0.895756E-02
	(−1.00745)	(0.882582)
BEDROOMS/ACRES	30991.9	2.43826
	(10.1627)	(1.00623)
COTTAGES/ACRES	45767.0	0.707381
	(4.04247)	(6.68135)
MILK/ACRES	837.285	−0.071913
	(21.1340)	(−1.47377)
SOIL1	289.986	0.085256
	(1.07736)	(0.784659)
SOIL2	447.601	0.169580
	(2.17045)	(2.11061)
SOIL3	342.133	0.125188
	(1.85922)	(1.73464)
SOIL4	−17.7765	−0.010473
	(−0.095308)	(−0.127354)
SOIL5	−403.345	−0.152597
	(−1.18479)	(−0.980049)
SOIL6	612.237	0.184909
	(2.50656)	(1.91875)
FD (Summer)	−36.7917	−0.047963
	(−0.455664)	(−1.68003)
FD (Winter)	17.5375	0.022348
	(0.730558)	(2.59075)
TEMP (Summer)	−113.941	−0.442380
	(−0.207969)	(−2.17610)
TEMP (Winter)	−133.051	0.294487
	(−0.322869)	(1.92768)
PRECIP (Summer)	0.316727	0.157462E-03
	(0.150734)	(0.155517)
PRECIP (Winter)	0.388410	0.158691E-03
	(0.329786)	(0.316864)

Table 3.3 *(Continued)*

Dependent variable	Price/acre	log (Price/acre)
SUN (Summer)	0.306286	0.219481E-03
	(0.070680)	(0.111423)
SUN (Winter)	-0.620128	0.132885E-03
	(-0.070295)	(0.031766)
REH (Summer)	-68.3107	-0.107184
	(-0.678215)	(-2.45004)
REH (Winter)	-10.6338	0.028819
	(-0.108262)	(0.710299)
WIND (Summer)	442.145	0.640366
	(0.476536)	(2.04126)
WIND (Winter)	-490.081	-0.459969
	(-0.976176)	(-2.89128)
Number of observations	405	405
R-squared	0.803762	0.595156
Fisher test for zero slopes	55.0015[1]	19.7411[1]
Breusch Pagan test for Heteroscedasticity	80.1029[1]	29.6323

Note Figures in parentheses are t-statistics.
[1] Significant at 1%

Discussion

The issue of primary interest relates to the significance of the climate variables. The hedonic price schedule given by the semi-log model is differentiated with respect to these variables to obtain the implicit price schedule. The implicit prices are evaluated at sample means and presented in Table 3.4. These implicit prices have been annuitized at 3 per cent per annum so that they refer to the annual stream of benefits associated with an additional unit of each climate variable.[10] Of the 12 climate variables, two are significant at the 1 per cent level of confidence, three are significant at the 5 per cent level of confidence and a further two are significant at the 10 per cent level of confidence.

It would be easy to blame the insignificance of the variables describing summertime and wintertime averages for precipitation and sunshine as due to the high correlations which exist between them (0.91 and 0.96 respectively). However, the correlation between the average number of frost days in summertime and wintertime is 0.91 while the correlation between summertime and wintertime averages for windspeed is 0.93, and in both of these cases the variables are significant at the 1 per cent level of significance for wintertime. An alternative explanation for the insignificance of the variables describing precipitation and sunshine is that they are close to their optimum values and that marginal changes in the levels matter little (also see below). The use of higher order terms for the 12 climate variables did not significantly increase the fit of the equation. This is unsurprising when one considers the lack of climatic extremes in the data set.

In several cases the marginal value of a climate variable depends upon the season and in some instances the direction of the impact is reversed. The use of both extreme measure and average values of the same climate variable appears justified: both the number of frost days and mean temperature are statistically significant. This appears to be a confirmation of what many commentators have suggested would be the case for climate change. Changes in extremes as well as averages are important (see for example Mearns et al, 1984).

What is more surprising is that an increase in the number of frost days during winter increases land values. This could be explained by stating that a cold snap during wintertime kills pests and vermin to the benefit of agricultural production.[11] Wind speeds might have been expected to have had a significant influence on land values insofar as strong winds are likely to cause erosion of the topsoil during the winter when the fields are bare. But because the full amenity value of climate to agriculture can only be inferred by joint consideration of the markets for land and labour, this limits the ability to place interpretations upon the sign of the coefficients in Table 3.4.

Notwithstanding the problems caused by the absence of a hedonic wage rate regression for farm workers, it would be tempting to present these values as a measure of the *value* of climate change's impact on England and Wales so far as agriculture is concerned. But it must be further acknowledged that whereas the implicit prices genuinely reflect the willingness of the farmer to pay for marginal changes in the level of climate variables, the values viewed from the perspective of society are different. The value of agricultural output and therefore land values is inflated due to the operation of the Common Agricultural Policy (CAP).[12]

One could be content with just calculating the implicit prices from the perspective of the farmer or alternatively one could attempt some kind of *ad hoc* adjustment to the land prices. One possibility, not attempted here, is to determine the impact of CAP interventions on the level of land prices using a time series econometric model. The model might then be used to simulate the removal of the CAP and the results used to adjust land prices in order to bring the values expressed by the farmer more into line with the values of society. The only published attempt to ascertain the impact of the CAP on UK land prices is the work of Traill (1979). This procedure, however, assumes that the effect of subsidies has an impact evenly across all climatic zones. It is possible that certain subsidies are targeted at areas with particular climates (eg as in subsidies to hill farmers). Insofar as the price of farmland reflects the subsidies given to those who work it, the amenity values of the climate variables might be distorted. An alternative way forward might be to include the amount of subsidy received by different farms as a separate explanatory variable.

Turning to the other non-climate variables, originally many hedonic studies were conducted using private or professional valuations rather than sale price data (Freeman 1993). Here the presence of both actual market data and professional valuations make it possible to test whether the intercept of the model is different for the two price measures (for simplicity, common slopes is a maintained hypothesis). Clearly, if the guide price is an unbiased predictor

Table 3.4 *The Impact of Climate Variables on Farmland Prices (Evaluated at sample means and annuitized at 3% per annum)*

Variable	Price (£)
Frost days (summer)	-3.80^3/frost day/acre
Frost days (winter)	$+1.77^1$/frost day/acre
Temperature (summer)	-35.06^2/°C/acre
Temperature (winter)	$+23.34^3$/°C/acre
Wind speed (summer)	$+50.76^2$/m/s/acre
Wind speed (winter)	-36.46^1/m/s/acre
Precipitation (summer)	$+0.01$/mm/acre
Precipitation (winter)	$+0.01$/mm/acre
Sunshine (summer)	$+0.02$/hour/acre
Sunshine (winter)	$+0.01$/hour/acre
Relative humidity (summer)	-8.50^2/percentage point/acre
Relative humidity (winter)	$+£2.28$/percentage point/acre

Note [1] Significant at 1%
[2] Significant at 5%
[3] Significant at 10%

of sale price then the coefficient on the variable 'private' should not differ significantly from zero. In reality the guide price appears to slightly overestimate the eventual sale price. Furthermore, there is no evidence that the error variances differ between observations which are based on the guide price rather than the true sale price.[13]

The coefficient on the variable indicating vacant possession is both positive and significant, illustrating that regulations governing tenancies hold rents beneath market values and reduce the value of the property to institutional investors. More specifically, the same property sold with vacant possession fetches more than double the price on the open market. The land quality grading system plays an uncertain role. Given the way that they are commonly presented in a sales prospectus, one might have anticipated that the coefficients on these variables would decline as land quality moves from Grade 1 to Grade 6. In fact, however, while the value of land declines as one moves from Grades 2 to 5, Grade 1 farmland is not the one valued most highly of all. The most valuable land appears somewhat perversely to be Grade 6. This underlines the fact that the grading system uses physical rather than economic criteria with which to classify land (see above). One explanation for the high value of Grade 6 land is that some of it might be being purchased with a view to the construction of houses on it.

CONCLUSIONS

This chapter has tested a new and alternative means of calculating the value of marginal changes in climate to agricultural production. The results uphold the findings of earlier analyses conducted in the US, India and Brazil in that amenity values for climate seem to be embedded in agricultural land prices.

Making more detailed comparisons between the studies is more difficult since there is a limit to how similar the amenity values for marginal changes in particular climate variables can be expected to be, given the very different climatic and economic conditions which prevail in these countries. There also is evidence that the use of seasonal values rather than annual averages is important and that the use of measures of climatic extremes can improve the fit of the hedonic land price regression. This is seen most clearly in the fact that even though average temperatures are included in the hedonic regression equation, the number of frost days is still significant.

There are unfortunately many caveats attached to the results which prevent them being combined with climate change scenarios and used to infer the impact of climate change on agriculture. In the first place, the value of climate to the farmer is not necessarily the value of climate to society by virtue of the CAP. Secondly, the technique cannot identify the impact of the CO_2 fertilization effect on productivity. Depending on what one believes to be the strength of this effect in the context of the field (rather than in the laboratory) the impact of climate change on agriculture inferred from hedonic studies may provide an over-pessimistic assessment. The potentially most important caveat, however, is one that has been overlooked so far. The implicit prices of climate amenities are measured only insofar as they are embedded in the hedonic land price schedule (and not in the wages of agricultural workers). Any future analysis should therefore attempt to determine whether land prices on their own contain all the information on the amenity value of climate or whether part is embedded in agricultural wage rates as well. A secondary objective ought to be the inclusion of farm-level subsidies into the hedonic price equation in order to see whether these have the effect of compensating farmers working in climatically less favoured areas.

The chapter also produces a set of other interesting findings unrelated to amenity values. It demonstrates that the guide price used by land surveyors is an unbiased guide to the eventual sale price. It demonstrates that tenanted farms are sold for significantly less than similar farmsteads with vacant possession. Finally, it demonstrates that farmland cannot be costlessly repackaged, in the sense that the size of a plot exerts a significant effect on the eventual sale price per acre and that the value of the structural attributes of land is not independent of plot size.

NOTES

1   Elevated concentrations of CO_2 have been shown to have a beneficial effect on plant growth in laboratory experiments. It is argued by some researchers that this fertilization effect may provide an offset against the potentially deleterious impacts of climate change on agriculture (eg Bolin et al, 1986). There is, however, continuing uncertainty over whether the large increases in the production of dry matter observed in laboratory conditions are indicative of what can be expected in open field conditions (eg Erickson, 1993). To the extent that hedonic analyses ignore the fertilization effect, they either exaggerate the negative impacts of climate change or understate the benefits.

2   For a general discussion on the theoretical basis for and empirical implementation of the hedonic technique see Palmquist (1991).

3   While it has been suggested to the author in private correspondence with the National Farmers' Union that wages paid to farm labourers do not vary much across the country, it is nevertheless true that they spend their working lives outdoors. They may not be indifferent to the climates in which they work.

4   Agents voluntarily submit details of sales to the journal and it is generally considered that the land prices quoted exceed those given in official statistics.

5   These were provided by the Climate Research Unit of the University of East Anglia under the auspices of the TIGER initiative.

6   Most tenants and their successors have security of tenure for up to three generations.

7   Note that the proportionality of the price of locational characteristics to farm size is a maintained assumption in what follows. Testing this assumption would require an additional 19 degrees of freedom.

8   Note that the assumption of normality is not upheld in Table 3.3. However, the consistency of the Ordinary Least Squares estimator does not depend upon the assumption of normality and the continued use of the t-test to test the significance of individual parameters and the use of the c^2 statistic to test combinations of parameters can be justified asymptotically (Greene, 1997). Furthermore, following the failure of the assumption of homoscedasticity in one of the models and the possible consequences of non-normality on tests for heteroscedasticity, all standard errors are calculated using White's method (White, 1980).

9   It is not possible to attempt a similar transformation for the right-hand side variables since many of them take zero values.

10 The use of a 3 per cent discount rate is suggested by the work of Lloyd et al (1991) in their examination of land prices in England and Wales.

11 There is a lugubrious old saying in eastern England that 'a warm winter makes for a fat graveyard'.

12 Even if one had estimates of the full implicit price of climate variables in order to move to an assessment of the full impact of climate change on agriculture, one would need to assume that the price of inputs and outputs remains constant.

13 This is confirmed using the Goldfeld Quandt test for homoscedasticity (Goldfeld and Quandt, 1965). The F statistic with 226 and 121 degrees of freedom was 1.11 with the critical value at the 5 per cent level of confidence equal to 1.32.

Chapter 4

IN SEARCH OF WARMER CLIMATES:
THE IMPACT OF CLIMATE CHANGE ON FLOWS OF
BRITISH TOURISTS

The preceding chapters have dealt with the amenity value of the British climate. But as British people now spend more time away from home, largely for the purposes of recreation, the climate of other countries may also be important to their welfare, if only for purely selfish reasons.[1] The rapid growth in international tourism is a reflection of greater leisure time (due in part to an ageing population) and a growth in real incomes. There is also evidence linking the growth of package holidays in the Mediterranean with reductions in cost caused by more fuel-efficient planes and with a decrease in the cost of accommodation (Perry and Ashton, 1994). In addition to the relative price of different locations, choice of destination is influenced by a desire to visit particular landscapes or sandy beaches for recreational purposes, motivated by a desire to explore or renew cultural ties between countries or to partake of the alleged health benefits of particular locations. Poor health has often been cited as a reason for making a journey (eg the remedial properties of hot spas, mountain air and coastal climates). Even until quite recently a tan was considered rather a 'healthy' thing to possess. Choice of destination is also heavily influenced by the image that a country has with regards to its political stability and crime rate (eg the recent poor publicity surrounding Florida following the murder of several British tourists).

But a major factor in choice of both destination and time of departure is climate. Indeed, when British tourists go abroad they are often described as being 'in search of warmer climates'. Tourists might be construed as making a decision to go abroad in order to gain some short-term climatic advantage. Certainly, in America retired people migrate south for the winter to Mexico, while in Australia they head north to the 'Gold Coast' resorts of Queensland. Both 'push' and 'pull' factors are clearly at work. While the importance of climate to both domestic and international tourism is not disputed, there are few studies which have explored the implications of climate change for international tourism even in a qualitative fashion (see Wall, 1992, for an exception).

The potential impact of climate change on tourism within Britain is dealt with by Smith (1991) who predicts than the tourist season may lengthen and tourist satisfaction may increase. But he makes no attempt to determine the changes in the overseas destinations of British tourists, changes in the number of holidays taken nor to value the costs or benefits from changes in the climate in terms of international tourism. The purpose of this chapter is to assess in a quantitative fashion the importance of climate as a determinant of choice of travel destination for British residents among a number of other possible factors including travel cost.[2]

From a purely strategic perspective it is of interest to examine how the numbers visiting different sites change as the climate changes. Many island economies in particular are heavily dependent on tourism and if climate is what tourists are seeking then climate change may have significant consequences for them. Furthermore, following the methodology outlined in this chapter, it is possible to compute a money-metric measure of how welfare changes as the attributes of a set of sites change (the welfare of tourists changes in the sense that more desirable climates may be brought closer to home). There is also a possibility that several low-lying island states may become 'unavailable' in the sense that they risk being inundated by rising sea levels. The methodology employed enables monetary values to be placed on this eventuality insofar as 'use' values are concerned. The methodology rests on the fact that different sites are characterized by different travel costs and accommodation costs. By observing differences in visitation rates it is possible to examine the rate at which individuals are willing to trade off higher money costs against desirable site attributes such as climate. The economic values of changes in both site quality and availability may be of interest to those seeking to compile an overall damage cost assessment of the effects of climate change.

CLIMATE AND TOURISM: THE ASSUMPTION OF PERFECT INFORMATION

The first information that potential holidaymakers encounter regarding the climate of a particular destination is through the travel company's brochure. The image of an attractive climate is cultivated in the mind of the consumer bounded by the requirement to remain within certain standards. There is obvious scope to be selective in the presentation of particular climate variables. The existence of warm temperatures in the Mediterranean during wintertime is emphasized, but the relatively higher rainfall (compared to London) is downplayed. Holiday tour operators frequently use the same 'blue sky' photographs in their summer and winter brochures. Even when climatic tables are given, there is a concern that the data are not readily understood and that some 'expert interpretation' is required.

The view that tourists were largely ignorant of, or were purposefully deceived about, the climate of their intended destinations (and therefore suffered disappointment as a result) and were moreover incapable of assessing for themselves raw climate data, led to the construction of indices which purported to objectively evaluate tourist potential of the climates of different countries. In the view of the contributors to this literature (eg Mieczkowski,

1985) the climatic resources of the world were not being fully used or, to put it another way, the inadequate provision of information resulted in market failure. In these indices different components of climate were subjectively weighted and placed on a labelled scale.

Perhaps as a result of the view that tourists do not have access to perfect information, there are no examples of taking a revealed preference approach to tourist flows in the literature. But are tourists really so lacking in information? Do they allow themselves to be persuaded by blue-sky photographs? Furthermore, the traveller has ready access to sources of high quality, low cost information which is independent of the travel company. This is in the form of countless travel guides, television programmes, daily weather reports for world capitals published in the newspaper as well as on television and radio and weather guides such as Pearce and Smith (1993). The main source of information, however, is surely from people who have already visited a particular destination.

Since the cost of obtaining this information is low relative to the cost of a typical package holiday, the tourist has every incentive to take advantage of it (Perry, 1993). It is therefore hard to conceive how travel companies could consistently mislead tourists with respect to what type of climate to expect, and the concerns of earlier researchers appear misplaced. As such, the revealed preference approach, which rests on the assumption of perfect knowledge concerning the attributes of different destinations, can be expected to work well, whereas much of the earlier literature looks like an anachronism. Furthermore, the revealed preference approach is capable of expressing in monetary terms the extent to which changes in the climate of different holiday destinations change welfare, which is a fundamental objective of this chapter.

THE POOLED TRAVEL COST MODEL

The travel cost method is most often used to estimate the value of individual recreation sites. In this method decisions on whether or not to visit particular sites are based on the various costs associated with travel to the site, and on the benefits derived from using it. Comparing different levels of visitation to a site from populations with different travel costs to reach that site, then allows the value of the site to be estimated.

The most straightforward form of the technique is the zonal travel cost method. In this method the area surrounding the site is divided into concentric circular zones, for which travel cost is estimated for a return trip; visitation rates are calculated for each zone (number of visits divided by population) and the rates are then regressed on travel cost and selected socio-economic variables. The observed total visitation represents one point on the demand curve for each zone; other points on the curve are estimated by assuming that visitors respond to an increase in admission price in the same way that they would respond to an increase in travel cost.

There are several potential sources of bias in the travel cost method, and the following list is not exhaustive. For one, the model relies on the assumptions that the demand structures are identical, and that preferences are the same in every respect other than is controlled for in the regression equation. In order for the full travel cost to be the correct variable for determining benefits, a visit to the site must be the sole purpose of the trip. Often, however, visits to one site may be combined with visits to other sites (whether for recreational or other purposes). Perhaps the single most intractable difficulty with the travel cost method lies in the valuation of the total time spent on travel, and the valuation of time spent on site. Thus, the value of several hours of each individual's time is sought. The zonal travel cost technique is reviewed in Freeman (1993).

The travel cost technique may also be used to estimate the value of a quality change at a site. The procedure invokes the assumption that the characteristic of interest is weakly complementary with a private good; in other words, the assumption is that the environmental quality variable is valued only if a trip is made. A theoretical model of the allocation of time and money spent visiting different sites and the consumption of other goods follows. The model is based largely on McConnell (1992) and Johansson (1987). Unlike in the majority of the travel cost literature, the time spent at particular sites is modelled as being a variable of choice and places particular emphasis on the valuation of site characteristics.

Assume that the individual derives utility from the number of visits to different destinations, the time spent on each visit, as well as from the consumption of a vector of other goods. The individual's utility function can be written as:

$$u = u(q, x, t, z)$$

where u is utility, q is a vector of consumption goods, x is a vector containing the number of visits made to each site, t is a vector detailing the time spent on each visit to site and z is a vector of site quality. The constraint attached to the choices made is as follows:

$$qp_q + xp = m + lp_w - tp_w - ax \ p_w$$

where p_q is a vector of prices, p is the (ticket) price of travel, m is unearned income, l is the amount of time available for work, p_w is the wage rate and a is the time required to visit a site. Thus the final term of the right-hand side of the equation represents the economic cost of time spent travelling. Associated with the solution to this problem of constrained maximization is the indirect utility function V (in which some constants are suppressed in order to simplify the notation):

$$V = V(p, z)$$

Employing Roy's theorem yields a set of demand equations for the number of visits to each site:

$$\frac{V_p}{\lambda} = -x(p, z)$$

where λ is the marginal utility of money (which is treated as a constant). Let p^0 represent the current price of travel and p^c a price so high that no trips are taken at all (p^c may be infinity and is often referred to as the 'choke' price). Integrating both sides with respect to p between the limits of p^c and p^0 gives the Consumer Surplus (CS) obtained from the site:

$$CS = \frac{V(p^c,z^0)}{\lambda} - \frac{V(p^0,z^0)}{\lambda} = -\int_{p^0}^{p^c} x(p,z^0)\ dp$$

Next, differentiating both sides with respect to z gives:

$$\frac{V_z(p^c,z^0)}{\lambda} - \frac{V_z(p^0,z^0)}{\lambda} = -\int_{p^0}^{p^c} x_z(p,z^0)\ dp$$

But given the assumption of weak complementarity between x and z (see Freeman, 1993):

$$\frac{V_z(p^c,z^0)}{\lambda} = 0$$

In other words, if it can be assumed that there exists a price so high at which the number of trips taken to the site falls to zero, then changes in the level of the site attribute z do not affect utility. Integration of this equation with respect to z gives the change in CS following a change in the level of site attribute z:

$$\Delta CS = -\int_{P_0}^{P_c} \left[x(p,z^0) - x(p,z^1) \right] dp$$

where z_0 and z_1 are the pre- and post-change level of site attribute. Even though several commodities may exhibit weak complementarity with environmental quality (eg flight costs and accommodation costs) all that is required in order to measure the value of changes in environmental quality is the demand curve for one of those commodities.

When many alternative sites are being studied, there may be substantial variation in qualities across sites. But while travel costs to the same site may differ between individuals (in the empirical application of the model described below they do not) the site quality is the same for everyone. Therefore all empirical models which attempt to incorporate site quality have involved some kind of simplification and as a consequence suffer limitations in their ability to characterize recreation demand accurately (Freeman, 1993). One approach (see eg Smith et al, 1986, and Caulkins et al, 1986) has been to pool all the observed visitation rates for the different sites and to estimate a single demand function in which the observed visitation rates are solely a function of the own price and quality variables:

$$x_j = x(p_j, z_j) \forall j$$

This model is referred to as the Pooled Travel Cost Model (PTCM) and is used in this study. Its fundamental weakness is that it predicts changes in the overall number of visits to a group of sites but does not allow for a reallocation of visits between different sites following a change in the price or quality attributes of alternatives. Moreover, it assumes that the coefficients on the own price and quality variables are the same across all sites. By contrast, reallocation effects are dealt with explicitly by the Random Utility Model (RUM) of choice approach to valuing site attributes.

With the RUM an individual chooses from a set of alternatives according to the utility which they provide. The indirect utility function V associated with a particular site j and choice-occasion is:

$$V_j\left(z_j, y - p_j\right)$$

where z_j is a vector of site attributes, y is income and p_j is the price of visiting the site. The indirect utility function contains a random error term which means that the choices made cannot be predicted with certainty but only with a given probability. The random error term reflects the existence of unobserved site characteristics and/or variations in taste between individuals. An individual visits a particular site k provided that:

$$V_k\left(z_k, y - p_k\right) + e_k > V_j\left(z_j, y - p_j\right) + e_j$$

$\forall$ sites $j \neq k$

If the random error terms are distributed as type I extreme value variates, then the probability of an individual i making choice j is given by:

$$Prob\left(Y_i = j\right) = \frac{e^{V_{ij}}}{\sum_{j=1}^{j=n} V_{ij}}$$

This is referred to as the Conditional Logit model. This model, however, has well known shortcomings of its own, one of which is the implicit assumption of the Independence of Irrelevant Alternatives (IIA). A further defect of the RUM is that it is incapable of predicting any possible change in the total number of tourist trips made following a change in site attributes: all that is predicted is how an exogenously determined number of trips is allocated between different destinations (eg see Bockstael et al, 1986). The only reason why the PTCM is preferred to the RUM in this instance is because the RUM is computationally difficult when the number of choices exceeds more than just a few.

DATA AND SPECIFICATION

Having outlined its theoretical underpinnings, this section describes the data sources for the variables used to estimate the PTCM as described by the PTCM equation on page 57. There is little existing literature to act as a guide to the appropriate specification of the model, so the analysis should be looked upon as probationary.

For the dependent variable, quarterly data on international travel by British residents are taken from the International Passenger Survey (IPS) for 1994.

Visits abroad for reasons other than holidaymaking (eg business trips) are excluded since they are not as responsive to climatic factors. The data set also contains the average return fare paid per person to each destination, average spending on items other than fares and the average duration of the stay. From the latter two variables it is possible to determine daily expenditure. While this is not the same as having a sterling price index for the cost of living relevant to tourists, it is quite clear that accommodation costs are an important consideration to the potential tourist and also that some countries are considerably more expensive to stay in than others. It was argued in the theoretical model that the amount of time required to travel to the country of interest had an opportunity cost attached to it such that, other things being equal, nearby resorts are preferred. The average time spent in transit is not available, so as a proxy the 'great circles' distance from London is used instead; this is the shortest distance which an aircraft could fly to reach a particular destination. Ordinary Mercator projection maps found in most atlases cannot be used to measure the great circles distance and instead an azimuthal equidistant projection map is required.[3]

Gross domestic product (GDP) per capita in US dollars, converted using purchasing power parity exchange rates, is taken from the UNDP (1995). Its inclusion in the data set reflects the belief that countries with higher GDP possess better tourist infrastructure. Furthermore, some tourists might be upset by visions of poverty and squalor which would greet them in many low-income countries. Population and population density are taken from the *Times World Atlas* (1992). Population proxies for the quantity of what might be called the 'cultural capital' that a particular country possesses (eg notable museums, sites of historical significance and buildings of architectural interest). Population density proxies for what might be referred to as 'natural capital' (eg unspoilt areas and environmental quality). These are unashamedly broad terms. It is anticipated that whereas the former will be positively related to tourism flows, the latter will negatively affect them.

The attraction of some countries clearly lies in the fact that they possess unspoilt sandy beaches fit for recreation. The total length of beaches found in different countries is available from a report by Delft Hydraulics (1990) and is added to the data set. Climate variables are taken from Pearce and Smith (1994). The climate of the country's capital city is taken since this is the most relevant for tourists (although there are arguments for producing a weighted average of several records to represent the climate of the larger climatically more diverse countries). Two variables are included as a description of the climate: averaged maximum daytime temperature and precipitation on a quarterly basis. The former is included in both a linear and quadratic fashion. Including both linear and quadratic terms allows temperature to exert both a positive and negative influence on visitation rates depending upon the current temperature. Finally, three dummy variables are included to represent the different quarters. The role of these variables is to demonstrate that differences in visitation rates can be ascribed to climate rather than any other seasonal factors such as statutory holidays. Many other variables might be expected to have an important influence on holiday destinations even though they are not included in the data set. Foremost among these are variables describing the level of personal security. This might in some future study be satisfactorily proxied by the inclusion of countries' respective murder rates. Sunshine too is omitted since it is not consistently collected for very many capitals. This is unfortunate in that many tourist destinations (such as Cyprus) are renowned for their sunny climate.

In total, 305 complete observations are available from 88 different countries. The variables contained in the data set are listed in Table 4.1, and the characteristics of the data set are examined in Table 4.2. The different countries represented in the data set are listed in Table 4.3.

Turning now to the functional specification of the model, the demand equation for the PTCM is modelled as:

$$\frac{VISITS_j^\lambda - 1}{\lambda} = \alpha_0 + \beta_1 FARE_j + \beta_2 GDP_j + \beta_3 POP_j + \beta_4 POPDEN_j +$$

$$\beta_5 BEACH_j + \beta_6 PDAY_j + \beta_7 DIST_j + \beta_8 TEMP_j + \beta_9 TEMP_j^2 +$$

$$\beta_{10} PRECIP_j + \beta_{11} Q1_j + \beta_{12} Q2_j + \beta_{13} Q3_j + e_j$$

where the subscript j refers to each different observation in the data set. Two special cases were considered: $\lambda = 1$ and $\lambda = 0$. These refer to the linear and semi-log models respectively.[4] In the linear model the impact on visitation rates of a change in the level of any variable is independent of the level of any other variable, whereas in the somewhat more plausible semi-log model this is not the case. β_1, being the coefficient on the own-price variable, is expected to be negative, as are the coefficients $\beta_4, \beta_6, \beta_7, \beta_9$ and β_{10}. In contrast the coefficients $\beta_2, \beta_3, \beta_5$ and β_8 are expected to be positive. There are no prior expectations regarding the sign of the coefficients β_{11}, β_{12} and β_{13}.

Table 4.1 *Definition of Variables Contained in the Travel Cost Data Set*

Variable	Definition
VISITS	Number of visits from the UK
FARE	Average cost of a return fare (£)
GDP	GDP per capita (1992 US$)
POP	Population
POPDEN	Population density (persons per km^2)
BEACH	Beach length (km)
PDAY	Cost of an extra day's stay (£)
DIST	Great circles distance from London to the capital (miles)
TEMP	Quarterly averaged maximum daytime temperature of the capital city (°C)
PRECIP	Quarterly precipitation in the capital city (mm)
Q1	Takes the value unity for the first quarter, zero otherwise
Q2	Takes the value unity for the second quarter, zero otherwise
Q3	Takes the value unity for the third quarter, zero otherwise

Table 4.2 *Characteristics of the Travel Cost Data Set*

Variable	Mean	Standard deviation	Minimum	Maximum
VISITS	85134.	0.26600E+06	235.0	0.2332E+07
FARE	225.58	148.47	17.00	818.0
GDP	10015.	6950.7	620.0	0.2376E+05
POP	0.4948E+08	0.1570E+09	6700.	0.1100E+10
POPDEN	395.83	1378.5	0.2455E-01	0.1350E+05
BEACH	131.03	368.09	0.0000	2970.
PDAY	37.397	15.774	9.287	181.0
DIST	3565.9	2523.3	199.0	0.1168E+05
TEMP	22.496	8.5632	−3.667	38.67
PRECIP	83.404	69.773	0.0000	403.3
Q1	0.23279	0.42330	0.0000	1.000
Q2	0.25902	0.43881	0.0000	1.000
Q3	0.26557	0.44236	0.0000	1.000

Note Number of observations = 305

Table 4.3 *Countries Included in the Travel Cost Data Set*

Anguilla	Djibouti	Italy	Puerto Rico
Antigua	Dominican Republic	Jamaica	Romania
Argentina	Ecuador	Japan	South Korea
Australia	Egypt	Jordan	Seychelles
Austria	Fiji	Kenya	Singapore
Azores/Madeira	Finland	Lebanon	Slovenia
Bahamas	France	Luxembourg	South Africa
Barbados	French Polynesia	Malaysia	Spain
Belgium	Gabon	Malta	Sri Lanka
Bermuda	Gambia	Mauritius	St Lucia
Bolivia	Germany	Mexico	Sweden
Brazil	Gibraltar	Monaco	Switzerland
Brunei	Greece	Morocco	Syria
Canada	Greenland	Nepal	Tanzania
Canaries	Grenada	Netherlands	Thailand
Cayman Islands	Hong Kong	New Caledonia	Trinidad
Chile	Hungary	New Zealand	Tunisia
China	Iceland	Norway	Turkey
Colombia	India	Pakistan	UAE
Cuba	Indonesia	Philippines	Uganda
Cyprus	Iran	Poland	US
Denmark	Israel	Portugal	Venezuela

RESULTS

Using the method described by Maddala (1977) it was found that the semi-log model (corresponding to the case $\lambda = 0$) was indeed most likely to have generated the observed data. The results of the semi-log regression analysis are displayed in Table 4.4. The coefficients can be interpreted as the proportionate

Table 4.4 *The Estimated Pooled Travel Cost Model*

Ordinary least squares regression	Dependent variable = Log VISITS
Observations = 305	Weights = ONE
Mean of left hand side = 0.9244884E+01	Standard deviation of
	Left hand side = 0.1987143E+01
Standard deviation of residuals = 0.1438978E+01	Sum of squares = 0.6025611E+03
R-squared = 0.4980400E+00	Adjusted R-squared = 0.4756157E+00
F[13, 291] = 0.2220981E+02	Probability value 0.0000000E+00
Log-likelihood = −0.5366101E+03	Restriction (ß=0)
	Log-l = −0.6417184E+03
Amemiya Probability	Akaike Information
Criterion = 0.3610558E+01	Criterion = 0.2165703E+01

ANOVA source	Variation	Degrees of freedom	Mean square
Regression	0.5978556E+03	13	0.4598889E+02
Residual	0.6025611E+03	291	0.2070657E+01
Total	0.1200417E+04	304	0.3948739E+01

Variable	Coefficient	Standard Error	t-ratio	Probability x
Constant	8.3472	0.5370	15.545	0.00000
FARE	−0.56164E-02	0.1003E-02	−5.602	0.00000
GDP	0.73447E-04	0.1539E-04	4.772	0.00000
POP	0.13699E-08	0.5751E-09	2.382	0.01785
POPDEN	−0.20062E-03	0.6318E-04	−3.175	0.00166
BEACH	0.14725E-02	0.2569E-03	5.732	0.00000
PDAY	−0.12633E-01	0.5436E-02	−2.324	0.02082
DIST	−0.76482E-04	0.5964E-04	−1.282	0.20069
TEMP	0.17252	0.4205E-01	4.103	0.00005
TEMPSQ	−0.29564E-02	0.1022E-02	−2.893	0.00410
PRECIP	−0.11087E-02	0.1342E-02	−0.826	0.40954
Q1	0.10804	0.2466	0.438	0.66162
Q2	−0.28269	0.2390	−1.183	0.23792
Q3	−0.13340	0.2448	−0.545	0.58617

change in the number of visits per unit change in the level of the dependent variable.

Overall the regression is highly significant and manages to explain almost 50 per cent of the variation in the log of observed visitation rates. Furthermore, the hypotheses put forward in the preceding section are all upheld. Not all visitation rates are well predicted, which is unsurprising given the importance of country-specific factors in determining choice of destination. As expected, the coefficient on the own-price variable 'FARE' is negative and highly significant, indicating that more expensive destinations generate fewer trips. It is also observed that countries with a higher GDP per capita are likely to generate more trips, as are more populous countries, but that countries with a lower population density are preferred. Countries with greater numbers of beaches are well-liked. The variable indicating the cost per day of visiting the different sites is negative and highly significant, indicating that the more expensive a country is to stay in, the more infrequently it is visited. The variable describing the great circles distance from London has the correct sign, but has only marginal significance. In part this may be because of the relatively high correlation between distance travelled and fare price.[5]

Turning to the climate variables, the coefficients on the linear and quadratic terms describing quarterly averaged maximum daytime temperature are positive and negative respectively, pointing to the existence of an 'optimal' maximum daytime temperature for tourism of around 29°C.[6] Precipitation, on the other hand, has a negative coefficient indicating that greater rainfall deters tourists, although not significantly so. This suggests that a different measurement concept such as rain-days might have been more appropriate or else that other omitted climate variables like hours of sunshine have biased the coefficient on rainfall. None of the dummy variables describing the time of departure is significant, implying that it is climate rather than other seasonal factors which explain observed visitation rates.

DISCUSSION

Using these results it is possible to examine the change in consumer surplus following a change in site attributes. The PTCM is also capable of answering the question 'Do British tourists have measurable use values for the low-lying islands?' Under some climate-change scenarios, the future facing these particular tourist destinations may not be a change in 'site quality' but instead elimination through inundation. These benefits are estimated only for British residents and many caveats apply, not least the assumptions made about the unimportance of the price and quality of substitute sites. The effect of this particular assumption is to make it appear that changes in the price and quality of alternative sites have no effect anywhere else, while in reality significant substitution between sites can be expected. Nevertheless, it is possible to use the model to predict the percentage change in the number of tourists visiting each country following a change in the attributes of the choice set as well as the ensuing change in consumer surplus and the overall worth of the site itself in terms of use values. These uses are illustrated for three different countries: Greece, Spain and the Seychelles. The first two are of particular interest since they are among the most popular tourist destinations for British people, while the latter consists of a group of islands, some of which are threatened by rising sea levels.[7] The impact of climate change on Greece and Spain is investigated using assumptions for the change in climate taken from the Meteorological Office general circulation model as reported in Houghton et al (1990). This model predicts a uniform increase of around 2°C for southern Europe (30°N–50°N) by the year 2030, complete with changes in seasonal precipitation patterns following 'business as usual' emission assumptions.

Table 4.5 illustrates the situation for Greece, where there is a lengthening and a flattening of the tourist season with tourist numbers almost unchanged. The first, second and fourth quarters show an increase in consumer surplus whereas the third quarter marks a sharp decline as maximum

Table **4.5** *The Impact of Global Climate Change on British Tourists Visiting Greece*

Quarter	Temperature (change °C)	Precipitation (% change)	Tourist numbers (% change)	Consumer surplus/deficit (£)
1	+2	+5	+16.0	+189,429
2	+2	–5	+3.9	+3,992,970
3	+2	–15	–2.8	–6,325,774
4	+2	–5	+11.1	+4,691,700
Total			+0.7	+2,548,325

Table **4.6** *The Impact of Global Climate Change on British Tourists Visiting Spain*

Quarter	Temperature (change °C)	Precipitation (% change)	Tourist numbers (% change)	Consumer surplus/deficit (£)
1	+2	+5	+9.4	+13,610,068
2	+2	–5	+6.3	+16,541,888
3	+2	–15	–0.5	–1,972,685
4	+2	–5	+16.7	+27,258,392
Total			+6.3	+55,437,663

Table **4.7** *The Impact of Climate Change on British Tourists Visiting the Seychelles*

Quarter	Temperature (change °C)	Tourist numbers (% change)	Consumer deficit (£)
1	unchanged	–100	–623,887
2	unchanged	–100	–1,204,152
3	unchanged	–100	–191,938
4	unchanged	–100	n/a
Total		–100	> 2,019,977

daytime temperatures pass well beyond their optimum level of 29°C. Overall, however, there is a small increase in consumer surplus of just over £2.5 million. Table 4.6 illustrates the situation for Spain, a country whose attraction lies in its climate, low population density and many miles of beaches. The results of climate change for Spain are qualitatively similar to those for Greece, but given Spain's lower prices and slightly cooler climate the beneficial effects of climate change on tourism are more pronounced. There are large gains for both tourist numbers and consumer surplus in the first, second and fourth quarters. In the third quarter there is a small decline in tourist numbers, but overall consumer surplus for trips to Spain increases by almost £55.5 million and the number of tourists visiting Spain goes up by more than 6 per cent.

Finally, in Table 4.7 the total consumer surplus arising from trips to the Seychelles is estimated as being slightly more than £2 million for the first three quarters. This is the 'use' value for this group of islands and the amount which would be lost if the whole group were inundated, which is an exaggerated proposition. In comparison with the gains from Spanish tourism, this sum seems very small, as it would be for most of the islands in the Indian and Pacific Oceans. The reason is that these islands are very small, distant, expensive and not visited much as a consequence. As a result they generate little consumer surplus. This means that these low-lying islands are not without value, but rather that their main value is likely to be in the form of existence rather than use values. The benefits to outsiders from preserving these islands lie in the vicarious consumption of their services through films, literature and the appreciation of their cultural heritage. The value of these services cannot be estimated through the travel cost technique, but might be assessed through the use of contingent valuation methods.

CONCLUSIONS

It has been demonstrated that quarterly climate variables are able to explain differences in flows of tourists. In particular, it is shown that British tourists are attracted to climates which deviate little from an averaged daytime maximum of 29°C. Furthermore, as the attributes of low cost (ie nearby) destinations are likely to improve following climate change, this is likely to result in a sizeable welfare gain to British tourists, even in the case of southern European countries like Spain and Greece. Both these countries, however, experience a lengthening and a flattening of the tourist season. In contrast, the losses experienced by the possible inundation of low-lying islands in the Indian and Pacific Oceans are likely to be small because these destinations are, at least to British residents, very expensive and consequently little visited. But it is important to stress that the values which this chapter seeks to estimate are use values and not total economic values, which may be much greater.

At an empirical level there is also further work to be done in terms of specifying the demand equation, including alternative measures for precipitation (such as rain days) and variables representing hours of sunshine and the degree of personal security. It would be interesting to examine the role of socio-economic factors such as age and income in explaining travel patterns. At a theoretical level reallocation effects are not well dealt with in the PTCM because the effects of changes in the quality (and price) of substitute sites are set to zero – a most unlikely representation of what would in fact happen. In the RUM, by contrast, the number of holidays remains unchanged irrespective of the quality of the experience provided by different destinations, so neither approach is entirely satisfactory.

NOTES

1   In 1994 UK residents made almost 40 million trips abroad (CSO, 1995).
2   Surprisingly, a survey of the economics and geographic literature failed to uncover any empirical analyses of climate as a determinant of holiday destination.

3   A computer programme made available through the University of Michigan to mea-
     sure the great circles distance between the world's capital cities can be found on the
     internet site: http://www.indo.com/distance/.
4   In contrast to the previous chapters, the left-hand side variables are not transformed
     into logarithms since some of them take negative values.
5   The correlation coefficient is 0.81.
6   No attempt should be made to compare the optimal temperatures for tourism with the
     optimal temperatures for the provision of human amenity described in Chapter 2. The
     ideal temperature for leisure and recreational pursuits and the ideal temperature for
     everyday living and working are likely to be profoundly different.
7   The Seychelles consists of over 90 small islands situated in the Indian Ocean. They have
     a tropical climate and have recently become well known as a tourist resort. Most of the
     islands are low-lying, but the largest island, Mahé, has hills rising to 3000 ft.

Chapter 5

THE AMENITY VALUE OF THE CLIMATE OF ITALY

By David Maddison and Andrea Bigano

Climate influences health and determines heating or cooling requirements, clothing and nutritional needs. Climate also delimits recreational and leisure opportunities and particular types of climate are known to promote a psychological sense of well-being. All this suggests that people ought to be interested in certain climates as well as explaining why those living in climatically different regions are likely to consume different patterns of marketed commodities: it is to compensate for, or alternatively take advantage of, particular sorts of climate.

But if at the same time as being free to adjust patterns of demand individuals are also freely able to select from differentiated localities then climate itself becomes a choice variable. The tendency will be for the costs and benefits associated with particular climates to become collateralized into property prices and wages. Households are attracted to regions offering preferred combinations of environmental amenities and this inward migration increases both house prices within those regions as well as depressing the wage rates in local labour markets. Thus across different cities there must generally exist both compensating wage and house price differentials. In such cases climate becomes a choice variable and the value of marginal changes in climate can be discerned from hedonic house and wage price regressions (eg Palmquist, 1991).

This chapter attempts to determine whether house prices in Italy are significantly affected by climate. Italy has a very diverse climate and this makes it a particularly suitable country in which to try out the hedonic approach to revealing the implicit price of climate variables. The work draws heavily on Bigano (1996) and considers only the existence of compensating differentials in the market for housing. Compensating differentials contained in wage rates are not considered since these are unavailable at the required level of regional disaggregation.[1]

Changes in climate amenity values may prove to be a considerable proportion of the overall impact of climate change on society, but it is a subset that has received surprisingly little attention in the climate change literature. Nonetheless, possessing a money-metric measure of the impact of changes in climate on welfare may prove useful, especially in terms of whether the costs of abatement are justified by the benefits in terms of damage avoided. The great advantage of the hedonic technique is that since one is using current day analogues for future climate change it can be presumed that long-run cost minimizing adaptation has already occurred. Apart from its strong theoretical underpinnings it is in this sense that the hedonic technique possesses particular advantages over rival methodologies.

The hedonic approach also highlights the fact that, although climate change is a global issue, its effects on people's welfare are likely to be different in different locations. Some regions may be negatively affected by an increase in temperature, whereas other areas may actually benefit from it. In the view of some commentators, climate change may be of benefit particularly for northern European countries like the UK (eg Mortimer, 1996) once suitable changes to lifestyles have occurred. Incidentally, Mortimer argued his case by pointing out that the UK was predicted to approach the type of climate which currently characterizes central Italy: a climate which is widely regarded as being in some sense 'optimal'. Of course this begs the question of what will happen to the climate of southern and central Italy if the enhanced greenhouse effect takes a grip, as the many subsequent replies to Mortimer's controversial article pointed out. The analysis presented below enables us to speculate on the effects of that eventuality.

Naturally there are some objections directed at the strong assumptions underlying the hedonic technique. It has been suggested that the technique could not be applied to the task of measuring the amenity value of climate because of the existence of significant costs of relocation compared to the benefits of particular types of climates. But this observation overlooks the fact that society has had a considerable length of time to respond to differences in climate; far longer in fact than the lifetime of any household. More importantly, it has been argued that since the climate variables are generally undeviating over relatively large distances and because of political boundaries, family ties, geographic and cultural differences, the assumption of a unified market for housing and labour becomes questionable. This assumption is particularly important in that if geographically segmented markets do exist then pooling of the data might lead to biased coefficients in the hedonic price equations since a single regression line is fitted to what are effectively two or more spline-functions (Straszheim, 1974). Fortunately, however, the validity of this assumption can be tested for.

STUDIES INTO THE AMENITY VALUE OF CLIMATE

In his review of the literature, Leary (1994) judged that the amount of money people are willing to pay to live and work in a better climate ranges from several hundreds to several thousands of dollars per year. He also argued that some climate characteristics can unambiguously be regarded as amenities and others as disamenities. All the studies which Leary considered, although focusing exclusively on US SMSAs, differed widely in the actual specification of the hedonic model and the choice of the explanatory variables, so that comparison of results is almost impossible. Indeed, climate attributes have hardly ever been the main focus of the analyses performed, the only exception being the seminal paper of Hoch and Drake (1974).

Hoch and Drake analysed three samples of worker categories using US Bureau of Labor Statistics microdata for 86 SMSAs. They assumed a perfectly elastic supply of land, thus preventing amenity effects from being capitalized into rents. Climate was specified in terms of precipitation and summer and winter temperature, the squares of the two latter terms, an interaction term for summer temperature and precipitation, wind speed, degree days, snowfall and the numbers of very hot (>90°F) and very cold (<32°F) days. The analysis was performed separately for three categories of employment. In the first, the coefficients on climate variables were significant and had the expected sign only as long as regional dummies were excluded, whereas in the other two they performed better. The only other explanatory variables were regional dummies, racial composition and urban size. No account was given of important site-specific characteristics like crime rates, pollution and other quality-of-life indicators.

The first attempt to estimate the effects of climate on both wages and house prices together is found in Roback (1982). She used microdata for the 98 largest American urban areas. Roback's analysis included snowfall, degree days, cloudy days and clear days. These variables are all highly significant and their coefficients have the expected sign in the wages regression, but performed poorly in the rental price regressions where only population growth, population density and unemployment rates were significant.

In his analysis Smith (1983) used real wages as the dependent variables and employed a cost of living index as a deflator. The focus of the analysis was the differential effects of job-specific and site-specific characteristics on wages for different industries, job categories and workers of differing ethnic composition. Climate, represented by mean annual sunshine hours, higher and lower temperatures, annual average wind speed and precipitation, was one of the site-specific characteristics considered. Only sunshine proved to have a significant (negative) effect on wages, indicating that it is regarded as an amenity.

Hoehn et al (1987) used microdata for 285 SMSAs to estimate separate hedonic regressions for wages and housing expenditures. They controlled for structural characteristics of houses and individual characteristics of workers. The same amenity variables were used in both equations and included coast proximity, crime rates, teacher–pupil ratios, total suspended particulates, visibility and six climate variables (sunshine, precipitation, humidity, wind speed, heating and cooling degree days). Among the climate variables, only sunshine was found unambiguously to be an amenity; that is, its coefficient displayed the same sign in both equations, whereas the other climate variables' net effect depended on the relative magnitude of their coefficients. In Blomquist et al (1988) the same data set was used to derive quality-of-life rankings for the metropolitan areas considered.

To the best of our knowledge, outside the US there is an almost complete lack of published climate amenity value studies.

Data Sources

The primary data source for this investigation into the amenity value of the Italian climate is *Il Sole 24 Ore del Lunedi* which is a leading financial news-paper based in Milan. The paper publishes an annual report on the quality of life in different Italian provinces. Several sets of indicators are used: general economics; job and business activities; public services and environmental quality; crime rates, demographic characteristics and leisure quality. Apart from reporting the raw data, the paper also computes an overall quality-of-life index by ranking regions according to the best performing province on each measure and then summing across the different measures to provide the final ranking. In 1996 the province enjoying the highest quality of life was judged to be Sondrio in the region of Lombardy: in 1997 it was Siena in Tuscany. Such a ranking procedure is of course entirely arbitrary.

For our purposes the most interesting variable in the published data set is property prices. The property price data refer to average house prices in the semi-central zone of the provincial capital and are measured in thousands of lira per square metre. Data are available for five years from 1991 to 1995 at the provincial level and there are data records for each of the 95 provinces in Italy.[2] The relevant variables are contained in Table 5.1.

Unemployment has frequently been included in hedonic house price analyses, with the majority of studies suggesting that lower house prices are

Table 5.1 *Definition of Variables Included in the Data Set*

PRICE	Property prices in the provincial capital (1000 lira per m^2)
NOTOCCUP	Unoccupied houses as a percentage of all houses
POPDEN	Population density (persons per km^2)
UNEMP	Unemployed persons as a percentage of the labour force
PRECIP	Annual precipitation (mm)
JANPRECIP	January precipitation (mm)
JULPRECIP	July precipitation (mm)
TEMP	Annually averaged mean temperature (°C)
JANTEMP	January temperature (°C)
JULTEMP	July temperature (°C)
CLOUD	Annually averaged fraction of cloudy days
JANCLOUD	Fraction of cloudy days during January
JULCLOUD	Fraction of cloudy days during July
NORTH	Unity if the province is in the north of Italy, zero otherwise
MIDDLE	Unity if the province is in the middle of Italy, zero otherwise
SOUTH	Unity if the province is in the south of Italy, zero otherwise
SICILY	Unity if the province is in Sicily, zero otherwise
SARDINIA	Unity if the province is in Sardinia, zero otherwise
DUM92	Unity if the observation is drawn from 1992, zero otherwise
DUM93	Unity if the observation is drawn from 1993, zero otherwise
DUM94	Unity if the observation is drawn from 1994, zero otherwise
DUM95	Unity if the observation is drawn from 1995, zero otherwise

Source Leemans and Cramer (1991) and *Il Sole 24 Ore del Lunedi* (1991, 1992, 1993, 1994 and 1995)

Table 5.2 *Characteristics of the 95 Italian Provinces*

Variable	Mean	Standard deviation	Minimum	Maximum
PRICE	2156.10526	793.44266	1100.00000	5800.00000
TEMP	12.50649	3.53297	1.10833	18.14170
PRECIP	1011.95789	352.34644	530.00000	1889.00000
CLOUD	0.49013	0.054159	0.39833	0.61417
JANPRECIP	84.97895	35.98669	32.00000	243.00000
JULPRECIP	44.91579	39.43607	2.00000	164.00000
JANTEMP	3.17053	4.04565	−7.10000	10.90000
JULTEMP	22.09579	3.59186	9.50000	26.20000
JANCLOUD	0.37421	0.051283	0.27000	0.48000
JULCLOUD	0.64821	0.080452	0.44000	0.79000
UNEMP	9.00122	6.49831	0.070000	32.40000
DUM92	0.20000	0.40042	0.00000	1.00000
DUM93	0.20000	0.40042	0.00000	1.00000
DUM94	0.20000	0.40042	0.00000	1.00000
DUM95	0.20000	0.40042	0.00000	1.00000
POPDEN	234.59684	319.46038	36.30000	2638.50000
NOTOCCUP	24.20327	10.37085	6.24561	76.02030
NORTH	0.46316	0.49917	0.00000	1.00000
MIDDLE	0.22105	0.41539	0.00000	1.00000
SOUTH	0.18947	0.39230	0.00000	1.00000
SICILY	0.084211	0.27800	0.00000	1.00000
SARDINIA	0.042105	0.20104	0.00000	1.00000

Source Leemans and Cramer (1991) and *Il Sole 24 Ore del Lunedi* (1991, 1992, 1993, 1994 and 1995)

Note Number of observations: 475

required to compensate for the reduced probability of finding work. In this sense unemployment is like a 'disamenity' and would be expected to have a negative coefficient in the hedonic house price equation (Todaro, 1969).

Population density is included in many hedonic house price regressions despite the fact that according to some (eg Steinnes and Fisher, 1974) population density is itself determined by house prices. Alternatively one might argue that the building of new homes is regulated to such an extent in Italy that population density and in particular the contemporaneous value of population density may be taken as exogenous. Population density is likely to serve as a proxy for many important, yet non-measured, variables such as pollution levels and the existence of good transport links. The dataset also includes the number of unoccupied houses (some of which may actually be uninhabitable) which may be taken as an approximate indicator of the presence of environmental disamenities in an area.

Choosing the best way of representing a fluctuating amenity is one of the major problems in hedonic models. Climate, for example, has numerous characteristics such as temperature, wind speed, sunshine, precipitation and snowfall, all of which can be represented in different ways. One can use monthly or annual averages, extreme values or indices like heating and cooling degree-days. Leary's (1994) survey shows how disparate the choices have been. The persistent lack of explanatory power displayed by some variables in the existing literature could be due to insufficient attention being paid to the specification of the climate variables.

In the following, two parallel analyses have been performed, the first considering just the annual averages and the second the values of the climate variables in January and July. This permits us to investigate whether, as might be expected, the amenity value of a marginal increase in the level of particular climate amenities depends on the season in which they fall. In addition the variables describing annually averaged precipitation, temperature and cloud cover are entered as both linear and quadratic variables, which permits the implicit price of climate variables to vary in sign depending upon the current level of these variables. This is shown to be highly significant.

The climate data are taken from Leemans and Cramer (1991). This database merges records drawn from a variety of published sources and, after various checks for quality and reliability, a terrestrial grid is created at the 0.5° level of resolution. Only mean temperature,[3] precipitation and cloudiness are included in the data set and these are reported as monthly averages. The data refer to the typical climate recorded over the period 1941 to 1961. Temperature data recorded in the original sources often incorporated adiabatic lapse rates. The presence of such factors implies that temperature values were corrected as if the station were located at sea level. Fortunately Leemans and Cramer report temperatures corrected to reflect the modal altitude of the grid square above sea level.

The climate of each province was described by using the data in the grid square in which the provincial capital was located. For some provinces (Ancona Brindisi, Lecce, Bari, Chieti Pescara, Palermo, Siracusa Trapani and Macerata) adjacent grid cells had to be used because the Leemans and Cramer dataset contains values only for those grid squares which are more than 60 per cent land.

In Italy topographic characteristics and the influence of the seas that surround the peninsula lead to a noticeable variation in the climate of different regions (Table 5.2). Cantu (1969–1981) discusses eight climatically quite different regions: the Alps; the Po Valley; the Northern Adriatic; the Central and Southern Adriatic; Liguria and Northern Tuscany; Thyrrenia; Calabria and Sicily; and Sardinia.

In Alpine Italy climate is similar to that of Switzerland or Austria although with heavier precipitation. Winter is bitterly cold and is generally the driest season. A climate similar to the one prevailing in Central Europe can be found in the Po valley, although this area is colder than Paris or London in midwinter. Summers are generally hot and quite humid. The Northern Adriatic region is characterized by the influence of the Bora, a very cold and gusty wind. The southern reaches of the Alps and the Appennini Ridge form a natural shield from the Northern influences, and temperatures in Liguria, Tuscany and Thyrrenia are generally higher than those prevailing at the same latitudes to the east of the Appennini. Here, a typical Mediterranean climate prevails with mild winters and warm, sunny and dry summers. The Appennini themselves, however, are generally wet for the greater part of the year and can experience heavy snowfall in winter. In Calabria, Sicily and Sardinia the weather is generally characterized by a very long, hot and sunny summer with a high degree of instability in the winter.

Finally, in order to guard against the possibility of segmented markets, a set of region-specific dummy variables was created. For this purpose Italy was divided into five different regions: northern Italy (Liguria, Friuli, Trentino, Veneto, Valle d'Aosta, Emilia Romagna, Lombardy and Piemonte); middle Italy (Abruzzo, Umbria, Marche, Tuscany and Lazio); southern Italy (Calabria, Basilicata, Puglia, Molise and Campagnia); Sardinia; and Sicily. Naturally, given the north-south climate gradient present in Italy, these dummy variables compete with the climate variables. It is, however, worth retaining these variables in the regressions which follow simply in order to demonstrate that the results which emerge are not the consequence of geographical segmentation of the housing market.

RESULTS

Linear regressions on property prices per-metre-squared proved disappointing and, using the method described by Maddala (1977), it became apparent that a semi-logarithmic model rather than a linear model was more likely to have generated the observed data. In what follows, therefore, the dependent variable was the log of price per-metre-squared. The regression analyses explain 33 per cent of the variation in per-metre-squared property prices when climate is specified in terms of annual means (see Table 5.3). When climate is specified as January and July means, the fit for the regression equation falls to 32 per cent. In both cases a random effects model is employed to account for the correlation of residuals when observations are drawn from the same province.

Considering first the specification involving the annual means and their squared values as descriptions of the climate, the majority of the climate variables are significant at the 5 per cent level of confidence. The pattern of positive coefficients on the linear terms and negative coefficients on the quadratic terms points to the existence of optimal values for annually averaged climate variables. The coefficients on unemployment, population density and the percentage of unoccupied houses are all statistically significant and, in the case of unemployment and the number of unoccupied houses, have the anticipated signs. None of the regional dummies is significant, not even at the 10 per cent level of significance. The dummy variables relating to the year of the survey are, as one might expect, highly significant and indicative of gradual nominal price increases in the housing market. Turning to the second equation, in which climate variables are specified in terms of their January and July averages, only two of the climate variables are statistically significant.

Evaluating the implicit price of climate variables at the sample averages in Table 5.4 leads to an interesting finding: none of the implicit prices is statistically significant taken either individually or jointly; not even at the 10 per cent level of significance.[4] This is at first glance difficult to explain given the fact that the majority of the climate variables were statistically significant at the 5 per cent level of significance and as a group significant at the 1 per cent level of significance. The explanation is that the sample averages are very close to the

Table 5.3 *The Results of the Hedonic Price Regressions*

Dependent variable	log (PRICE)	log (PRICE)
PRECIP	0.869861E-03[3]	
TEMP	0.089141[2]	
CLOUD	32.9639[2]	
PRECIPSQ	−0.471929E-06[2]	
TEMPSQ	−0.312669E-02	
CLOUDSQ	−33.7396[2]	
JANPRECIP		−0.140665E-02
JULPRECIP		−0.741696E-02[1]
JANTEMP		0.038307
JULTEMP		−0.038434
JANCLOUD		0.615168
JULCLOUD		−4.17079[2]
UNEMP	−0.012614[1]	−0.012497[1]
POPDEN	0.273334E-03[1]	0.264922E-03[1]
NOTOCCUP	−0.568254E-02[2]	−0.753525E-02[1]
MIDDLE	−0.089951	0.015411
SOUTH	−0.056920	−0.085708
SICILY	0.262479	0.079204
SARDINIA	0.026644	−0.098941
DUM92	0.152080[1]	0.151658[1]
DUM93	0.200289[1]	0.199423[1]
DUM94	0.232742[1]	0.231712[1]
DUM95	0.305138[1]	0.304472[1]
CONST	−0.066873	11.3405[1]
Number of observations	475	475
R-squared	0.333627	0.324482

[1] Significant at 1%
[2] Significant at 5%
[3] Significant at 10%

Table 5.4 *The Annuitized Implicit Prices for Climate Variables*
(Evaluated at 1995 Sample Means)

Variable	Price (lira)
Precipitation (annual)	−10/mm/m²/year
Temperature (annual)	+1312/°C/m²/year
Cloudiness (annual)	−120/percentage point of cloud cover/m²/year
Precipitation (Jan)	−168/mm/m²/year
Precipitation (Jul)	−887[1]/mm/m²/year
Temperature (Jan)	+4583/°C/m²/year
Temperature (Jul)	−4598/°C/m²/year
Cloudiness (Jan)	+736/percentage point of cloud cover/m²/year
Cloudiness (Jul)	−4990[2]/percentage point of cloud cover/m²/year

Note also that a 5 per cent discount rate was used to annuitize the purchase price.
[1] Significant at 1%
[2] Significant at 5%

Table 5.5 *The Optimal Value of Climate Variables*

Climate variable	Estimated optimum	Sample average
Annual precipitation (mm)	922[1]	1013
Annual mean temperature (°C)	14.3[1]	12.5
Fraction of cloudy days	0.49[1]	0.49

[1] Significant at 1%

climate optima and a marginal change does not lead to a significant change in amenity values. This explanation is borne out by calculating the 'optimal' climates suggested by the regression results and comparing them to the sample means (see Table 5.5).

Turning to the second specification in which January and July averages are used, it can be seen that the majority of the implicit prices do not differ significantly from zero, even at the 10 per cent level of significance. This is likely to be caused by the high degree of collinearity between January and July averages.[5] Nevertheless two of the climate variables (precipitation and cloud cover in July) are statistically significant at the 1 per cent and 5 per cent levels of significance respectively. More specifically, both precipitation and cloud cover in July are viewed as a disamenity.

Even if the use of monthly averages does not result in statistically significant implicit prices it is still possible to test whether the marginal value of climate variables differs between the seasons. The test for equality of the January and July coefficients can be rejected at the 5 per cent level of significance for both precipitation and the degree of cloud cover. Precipitation is more of a disamenity if it occurs during July and cloud cover is more of a disamenity if it occurs during July. It also appears that the value of an extra degree centigrade is greater in January than in July (indeed, the value is actually negative if the temperature is increased during July). Unfortunately, the difference in the temperature coefficients is statistically significant only at the 10 per cent level of significance.

Taken together, these results strongly support the use of the hedonic methodology for the purposes of determining the amenity value of climate. The empirical estimates of the implicit price of climate variables are plausibly signed and reasonable in terms of magnitude. The analysis points to the existence of climatic optima that are close to the sample averages of a country whose climate is widely hailed as optimal. The analysis also demonstrates, as one might expect, that the amenity value of climate variables differs across the seasons.

CONCLUSIONS

The main conclusion of the analysis is that there is considerable empirical support for the hypothesis that information on the amenity value of climate is contained in property prices. This is the case whether the hedonic price equations are specified in terms of annual values or in terms of monthly values. That these findings emerge from models with a large number of region-specific dummy variables is all the more remarkable since one would have expected these dummies to compete with the climate variables. It also appears that the marginal willingness to pay for additional climate amenities depends on the season into which they fall and that the current sample averages for climate are very close to the climatic optimum. In this sense, any variation in the current climate of Italy is likely to cause a reduction in amenity values. On the other hand, since most of Italy is currently at the optimum, marginal changes in climate variables are likely to have only a very low value.

Despite the encouraging nature of the results the following qualifications are in order. First, the climate database did not include some relevant variables like wind speed and humidity. Second, the choice of explanatory variables is, as always, arbitrary, reflecting what was available rather than any other consideration. Finally, a second set of hedonic schedules for wages should be estimated in order to attain a complete measure of the implicit prices of amenity variables. Unfortunately this has not been possible because of the lack of suitable data.

NOTES

1   The amenity value of climate variables would be wholly reflected in property prices to the extent that climate is an unproductive factor for firms and land is not required as a factor of production (see Roback, 1982). The amenity value of climate would also be wholly reflected in the market value of land if institutional reasons prevent wage rates from differing between locations. This is likely to be the case in Italy.

2   Seven new provinces have recently been created but the database remained with the original set for the sake of comparison.

3   Mean temperature is defined as the average of maximum daytime temperature and minimum night-time temperature.

4   To determine the payment actually made by each household would require that these prices be multiplied by the floor space of each particular dwelling.

5   In particular, the correlation coefficient between the January and July temperatures is 0.83.

Chapter 6

THE EFFECTS OF CLIMATE ON WELFARE AND WELL-BEING IN RUSSIA[1]

By Paul Frijters and Bernard Van Praag

It is well known that differences in climate affect the quality of life. Climate differences are informally recognized by multinationals and states extending over several geographic areas which supplement wages by climate allowances. However, little has been written on the impact of climate differences on well-being, a notable exception being the study by Blomquist, Berger and Hoehn (1988), who assessed the monetary value of amenities including some climate variables.

A second exception is the approach outlined in Van Praag (1988), which departed from data from a European survey in which respondents were asked to evaluate their own income level in terms of 'good' and 'bad'. It was assumed that individuals who gave the same answer enjoyed an equal level of *welfare*. This allowed Van Praag to identify the monetary value to respondents of precipitation, humidity and temperature. This approach has been refined further and is used here. We also apply the same methodology to another subjective measure, namely the satisfaction with life as a whole, which we call *well-being*, as developed by Cantril (1965). Thus we obtain and compare estimates of the effect of climate on both *welfare* (say satisfaction with income) and *well-being*.

After estimating the effects of climate on both welfare and well-being in Russia and computing the concomitant equivalence scales for various geographic sites, we then consider different scenarios for climate change and predict the subsequent increase or decrease in the climate costs for individual Russian households.

METHODOLOGY

We want to assess the monetary value of differences in climate across regions by measuring the amount of extra income respondents in one climate need in order to be equally well-off as respondents in another climate. By comparing individuals in different regions, we can estimate the influence of each different climate variable on welfare. More formally, let us say that welfare U is a function of two variables, current income y_c and climate C (or any other amenity). We have:

$$U = U(y_c, C)$$

The monetary value of a climate change to an individual can be calculated by changing the climate conditions of an individual (by $\_c$), and asking how much income would have to change (with $\_y$) in order to keep welfare constant. This can be calculated as:

$$0 = U_y \_y + U_c \_c$$

where the derivative of U to y is denoted by U_y and the derivative of U to C by U_c.

The shadow price of a climate change is now:

$$\_y = -\_c \times U_c / U_y$$

If income itself depends on climate, which is frequently the case as companies and governments already partly compensate individuals for climate conditions, and if prices in a region (denoted by p_c) also depend on climate, for total compensation there has to hold that:

$$0 = \left[\_y + \_c \times dy / dc \right] U_y + \_c U_p \times dp_c / dc + \_c U_c$$

which yields a shadow price of:

$$\_y = -\_c \times \left[U_c / U_y + dy / dc + dp_c / dc \times U_p / U_y \right]$$

For an empirical analysis we do not need to know the actual function U, but only the indifference surfaces (points where welfare is the same). This means that we have to identify individuals who enjoy the same welfare level, in order to assess the shadow price of climate conditions for that welfare level. Herein lies the rub: how to identify persons at the same welfare level? Roback (1982, 1988) and Blomquist et al (1988) circumvent this question by simply postulating that all households with identical wage-earning capacities are at the same welfare level, which in turn is based on the assumptions of perfect mobility and that households have identical utility functions. As a result of these assumptions, differences in welfare cannot exist between regions as that would induce migration towards better-off regions. These assumptions are improbable for Russia, where mobility is low. Another drawback of this method is that the shadow prices of climate found in Blomquist et al, and in more recent hedonic pricing studies such as Maddison and Bigano (1997), include both the pleasure costs of climate and the monetary costs of climate, but cannot distinguish between them. The empirical problem is that the monetary costs of climate include unobserved price differences in non-traded goods. Consider, for instance, the price of the non-traded commodity 'feeling warm'. If an individual lives in a very cold climate, he will need to burn more fuel, will need more clothes and may need adapted transportation to stay warm. This will increase the price of staying warm. To find the real price of staying warm would need a very detailed and reliable set of prices and household expenditure data per region, which are not available. This makes it impossible to get at the monetary costs of climate in Russia via an analysis of the effect of climate on incomes and observed expenditures on marketed goods such as housing prices.

Another popular approach for measuring shadow prices, contingent valuation, avoids the problem of measuring welfare as well. Contingent valuation studies ask respondents how much they would be willing to pay for a particular change in circumstances, such as climate. People are simply asked to report the _y which keeps them at the same welfare level: respondents are asked to solve for _y:

$$U(y,C) = U(y+\_y+\_c \times dy / dc, C+\_c)$$

The main assumption that has to be made for contingent valuation to work, assuming for the sake of simplicity that respondents try their best to answer the question honestly, which is often dubious (see Hanemann, 1994, and Diamond and Hausman, 1994), is that respondents need to know what effect each possible climate change has upon their welfare.

Instead of using indirect methods to identify individuals at the same welfare level, we asked repondents directly about their present level of welfare, so that we could compute the shadow price of current climate circumstances by comparing individuals in different climates who gave the same answers. That is, we use a subjective utility concept. We shall look at the two different concepts of welfare and well-being in detail. The first stands for a narrow concept, say satisfaction derived from income or monetary welfare. The second concept is a much wider concept and stands for satisfaction with life as a whole. The two concepts are each measured by a separate measurement instrument.[2]

The first concept, introduced by Van Praag (1971), is expressed using the income-evaluation question (IEQ). This is a question module in which the respondent is asked to qualify five household income levels.[3] The IEQ runs as follows:

'While keeping prices constant, what before-tax total monthly income would you consider for your family as: very bad, bad, not good not bad, good, very good.'

The five answers of individual i are denoted by c_{ij}. Their empirical log-mean and variance are:

$$\sigma_i^2 = \frac{1}{4} \bullet \sum_{j=1}^{5} \left(\mu_i - \ln(c_{ij}) \right)^2 \text{, and } \mu_i = \frac{1}{5} \bullet \sum_{j=1}^{5} \ln(c_{ij}) \text{ respectively.}$$

To enable us to use this attitude question for welfare comparisons, we make the crucial assumption that all households in a language community attach the same verbal label to the same welfare level. Van Praag (1991, 1994) researched whether it is true that households attach the same meaning to verbal labels in a value-free context: repondents were asked to translate verbal labels to a point on a line of fixed length. It turned out that there was a remarkable uniformity in the responses: not only did respondents use the whole line for their answers, they also displayed a tendency to use intervals of the same length between verbal labels. The exercise was repeated by asking respondents to translate the verbal labels to numbers on a (0,1)-scale, with the same result of equal partition, that is $U_{\text{very bad}}$ » 1/10, U_{bad} » 3/10, etc. In that case, it is found that the verbal labels translated into numbers on a (0,1)-scale are well described by:

$$U_j = N\big((\ln(c_j) - \mu)/\sigma\big)(j = 1,...,5), \text{ where } N(.)$$

is the standard normal distribution function. In this study we consider μ as a want parameter and $(\ln(y)-\mu)$ as an ordinal welfare index, as σ is taken to be a constant.[4] So we only assume ordinal interpersonal welfare comparibility in the sense of Sen (1976) – see also Parducci (1995).

The individual parameter μ has been shown to be well-explained by age, family size, family income and other personal variables (Van Praag (1971), Hagenaars (1986), Van Praag and Flik (1992)). More precisely, empirical relationships like:

$$\mu_i(y_{ic}, fs_i, age_i) = \beta_0 + \beta_1 \ln(y_{ic}) + \beta_2 \ln(fi_i) + \beta_3 \ln(age_i) +$$
$$\beta_4 \ln^2(age_i) + \beta' AC_i + \varepsilon_i \qquad 1$$

have been found to hold in each study, not only for Dutch data but also for data of many other countries and over time. In equation (1), y_{ic} denotes the current household income of respondent i, fs_i denotes the family size, age_i denotes the age of the respondent; $\mu_i(y_{ic}, fs_i, age_i)$ is replaced by μ_i or $\mu(y_i)$ whenever this is unlikely to lead to confusion. $\beta' AC$ stands for a linear combination of climate variables. $\varepsilon_i \sim N(0, s^2)$ denotes the error term.

The welfare parameter μ_i was found to depend on the current income of the individual i. It follows that individuals with different current incomes evaluate a specific income level differently. This phenomenon, embodied in β_1, is known as preference drift (called the 'hedonic treadmill' by Brickman and Campbell (1971) in a more generalized context). It is an empirical expression of the notion that welfare functions are evaluated relative to current circumstances both inside and outside of households (see Van der Stadt et al, 1985).

It follows then that we can estimate how welfare, understood as a function of the extent to which income meets financial needs (that is, as a function of the welfare index $\ln(y_{ic})-\mu_{ic}$), varies with climate conditions. As μ depends on climate, we can calculate the amount of money needed to compensate households for different climate conditions. Such an equivalence scale is derived by comparing the income an individual needs to enjoy a specific welfare level in one set of circumstances to the income necessary for the same welfare level under a chosen set of reference circumstances. We describe the reference situation as $(\ln(y_{0c}), \mu(\ln(y_{0c})))$. By equating welfare levels and solving for current income, y_{ic}, we get:

$$\ln(y_{0c}) - \mu(y_{0c}) = \ln(y_{ic}) - \mu(y_{ic})$$

which yields after substitution of μ by equation (1)

$$\ln\left(\frac{y_{ic}}{y_{0c}}\right) = \frac{\beta_2 \bullet \ln\left(\frac{fs_i}{fs_0}\right) + \beta_3 \bullet \ln\left(\frac{age_i}{age_0}\right) + \beta_4 \ln^2(age_i)}{1 - \beta_1} -$$
$$\frac{\beta_4 \bullet \ln^2(age_0) + \beta' \bullet (C_i - C_0)}{1 - \beta_1}$$

2

We notice that this ratio or equivalence scale does not depend on the welfare level. Hence it is 'independent of base' (see Blackorby and Donaldson, 1991).

Because this method yields an estimate of the amount of money needed to compensate for different conditions, we can interpret this equivalence scale as measuring the cost differences related to different conditions. This way of arriving at equivalence scales is called after the place where the method was first developed and is known as the Leyden approach.

Now we turn to our concept of well-being, which is much broader than the financial satisfaction concept defined above and is familiar from socio-psychological literature. It is a satisfaction-with-life question developed by Cantril (1965):[5]

'On a scale from 1 to 10, whereby 1 stands for very unsatisfied and 10 stands for perfectly satisfied, how would you rate your life as a whole?'

The answers obtained, denoted by V_i, are numbers on a (1, 10)-scale. An equivalence scale can be calculated by comparing V-values or by comparing monotonic transformations of V-values. Both methods yield the same equivalence scales. Hence we looked at the compensation needed in terms of money to keep somebody at a constant level of well-being in spite of a change in his other variables, for example, fs or age. If climate variables have an effect on well-being as measured by the Cantril question, then we can also derive climate equivalence scales with respect to well-being. Our first expression of this method translates the Cantril-answers onto a $(-¥,¥)$ scale. Increasing the similarity with the Leyden method, the variable V^*_i is defined as:

$$V^*_i = N^{-1}\left(\frac{-0.5 + V_i}{10}; 0; 1\right)$$

We can now proceed by assuming that V^* is generated by:

$$V^*_i = v_0 + v_1 \times \ln(y_{ic}) + v_2 \times \ln(fs_i) + v_3 \times \ln(age_i) +$$
$$v_4 \times \ln^2(age_i) + v¢ \times C_i + \varepsilon^*_i$$

where ε^*_i denotes the error term. An equivalence scale is constructed by holding the level V^*_0 of well-being constant and by compensating changes in climate conditions by changes in income. It yields a similar expression as in equation (2), albeit with different coefficients. For the structural part we get:

$$V^*_i = V^*_0$$

which yields:[6]

$$\ln\left(\frac{y_{ic}}{y_{0c}}\right) = \frac{v_2 \bullet \ln\left(\frac{fs_0}{fs_i}\right) + v_3 \bullet \ln\left(\frac{age_0}{age_i}\right) + v_4 \bullet \ln^2(age_0)}{v_1}$$
$$- \frac{v_4 \bullet \ln^2(age_i) + v' \bullet (C_0 - C_i)}{v_1}$$

3

The resulting equivalence scale gives an estimate of the money-equivalent of the effect of different conditions on well-being and is called the well-being-scale. It includes both the monetary costs of climate and the pleasure value of climate. Notice that the scale is again independent of base. An alternative way to derive equivalence scales from the Cantril-question assumes that the answers V_i are generated by an ordered probit equation. The results appear to be similar although less significant (see Table 6.3).

Expanding on the existing methodology, we considered the bias that our estimates would have if the incomes of respondents vary with climate conditions, as we know to be the case in Russia. To see how this can be remedied, consider what effect climate now has on income:

$$\ln(y_{ic}) = \theta_0 + \theta'AC_i + \theta_1 \phi X_i + \varepsilon y_i$$

4

where C stands for a vector of climate variables; X stands for the variables used both for μ, and also education, 19 industry dummies and 12 dummies denoting several types of organization, like farms, public bodies and private companies.

After estimating (4) for the climate variables used to estimate μ, we can calculate the full effect of climate on welfare and well-being by deflating income for its climate specific component:

$$\ln(y_{non-climate}) = \ln(y_{ic}) - \theta'AC_i$$

5

This measure of income was then used to correct household income for the way in which the labour market already compensates individuals for different climate conditions. The total effects of climate on welfare and well-being can be found by adding a term to the effects we find without a climate-corrected-income. Thus we can add $\beta_i\theta'AC_i$ to the estimated climate effects of μ and $v_i\theta'AC_i$ to the climate effects of V^*_i.

Before turning to the empirical work, we will address the issue of precisely which climate costs we picked up.

By looking at the amount of money households need to reach a given level of welfare, we are measuring the long-term costs induced by climate. This includes the effect of climate on the price of non-traded goods such as the price-increasing effect of cold on the price of 'staying warm' via the increased need for heating or the price effect of climate on 'health' via medical costs. The Leyden method also captures the effect of climate on the price of traded goods (eg the effect of rain on agricultural prices and housing prices). Thus, by indirectly looking at the effect of climate on total household needs, we measure in a reduced form all the costs of climate at once. As prices or levels of financial need may vary across regions for other than climatic reasons, such as through the effect of differences in political and economic systems upon prices and production capabilities,[7] we add a limited number of regional dummies to pick up systematic non-climate differences between regions. Bias may also come in if respondents do not answer the IEQ correctly, that is if they have no firm idea as to what income they would need to realize a welfare level different to the present. The IEQ suffers from this problem to a lesser degree than other hypothetical questions such as used in contingent valuation, for two reasons. Firstly, we do not ask that the respondent is explicitly aware of the effects of climate on welfare, nor that he attempts to give a money translation of climate changes. We only ask that he knows how much money he needs to reach a certain welfare level in his present circumstances. Secondly, we take direct account of the fact that welfare depends on the reference position (via the preference drift phenomenon). Similarly, the Cantril-question is not hypothetical in any way: it is merely rather volatile, with evaluations of one's whole life being dependent on present mood and subject to random variations in interpretation so that many observations are needed to find the structural parameters. It is also possible that regional differences in culture and interpretation of the Cantril question would bias our results, although they were not picked up by the limited number of regional dummies we used (see Appendix 2).

DATA SET AND EMPIRICAL ANALYSIS

This study uses the first two waves of the Russian National Panel data set, that is the data of a panel of 3727 households who were interviewed in 1993 and 1994. After deletion of cases with missing values, 2508 observations of the first wave and 1904 observations of the second wave were aggregated.[8]

As a first step, we will compare the results of the Leyden-method for the 1993–94 Russian data set with the results that were obtained and reported by Van Praag in 1988 for a large European data set collected in 1979.[9] In that analysis, no account was taken of climate effects on income.

In Table 6.1 we present the estimated coefficient values of the μ-equation for Western Europe 1979 and Russia 1993–94, suppressing the country-specific effects for non-climate variables and a 1994 intercept dummy for our Russian surveys.

The European data set consists of surveys in the Netherlands, the UK, Denmark, France, Belgium, Italy, Germany and Ireland. Temperature is defined

Table 6.1 *Climate-μ Regressions for 1979 and 1993–94: The Amount of Income Needed to Reach a Fixed Welfare Level under Varying Climatic Circumstances*

	Europe 1979[1]	Russia 1993–94[1]
Constant	6.94	2.17
	(20.6)	(2.1)
Family size	0.11	0.22
	(10.8)	(12.5)
Family income	0.57	0.65
	(57.0)	(60.2)
Ln(TEMPERATURE)	–0.15	–0.72
	(5.0)	(4.7)
Ln(HUMIDITY)	–0.41	0.32
	(6.8)	(1.2)
Ln(PRECIPITATION)	–0.10	0.16
	(10.0)	(4.7)
	$R^2=0.6584$	$R^2=0.7254$
	N=13428	N=4412

[1]   Absolute t-values between parentheses.

as the logarithm of the average temperature in a year for the European data set and equals the logarithm of average temperature plus 50 for the Russian data set. This difference in definition is unavoidable because of the colder climate in Russia and should only influence the size of the temperature coefficient, not its sign. It does mean that the actual effect of a temperature change differs much less than the coefficients would suggest. Humidity is defined as the logarithm of the average humidity in a year. Precipitation is the logarithm of the average annual precipitation. The Europe 1979 figures are borrowed from Van Praag (1988).

The differences between the relationship between climate and costs in Russia and Europe are reflected in Table 6.1 by the difference in the climate cost structure, the coefficients and the significance of the climate variables. Bearing in mind that a higher value of μ implies a lower standard of welfare when income is constant, it appears that higher temperatures are preferred in both areas, and that humidity does not increase the financial welfare of Russian households, contrary to the situation in Western Europe. The greatest difference is with regard to the effect of precipitation, which is generally seen as positive in Western Europe whereas in Russia a lack of precipitation does not entail higher costs but lower costs. This may well reflect the fact that in Russia the lower average temperatures lead to lower levels of evaporation and hence a reduced need for precipitation. A lack of rain would thus be less of a problem in Russia than in Europe. Similarly, it may be the higher temperatures in Europe which lead to a preference for high levels of humidity ('wetness'). This suggests a set of interactions to be important.

It may thus be that there are more relevant aspects of climate not included in the earlier analysis. We have a whole list of variables for Russia, which were not available in Van Praag (1988). Moreover, there is more variation in climate in Russia than in western Europe.[10]

Turning to the full analysis of climate costs in Russia in 1993–94, we have 13 different climate variables to work with, including the average temperature in January and July (JANTEMP and JULTEMP), the average annual temperature (TEMPAV), the difference between maximum and minimum temperature in one calendar year (TEMPDIF), the average level of annual precipitation (PREC), the average amount of precipitation in the summer and winter (SUMMERPREC and WINTERPREC), the number of raindays a year (RAINDAYS), the hours of sunshine a year (SUNHOURS), average wind speed a year (WIND), average wind speed in January (JANWIND) and the height of the region above sea level (HEIGHT). The precise definition of the variables is controlled in Appendix 2.

It can be expected that the effect of climate on welfare and well-being is the result of several interactions. Some interaction terms therefore were also added. As we could distinguish only 35 different climate regions in the Russian 1994 data set, the number of climate variables included had to be restricted.

The only strong a priori expectation we had about the relationship between climate on the one hand and cost of living and well-being in Russia on the other is that we expected the extremely low temperatures to lead to a reduction in welfare and well-being. In the absence of other expectations, climate variables were selected on the basis of best fit: the model that explained the most variance was selected for both the Leyden concept and the Cantril concept. Then the wage-regression was run, for which the same variables were used as those selected for welfare, which turned out to fit reasonably well for wages. The full procedure is described in Appendix 2. The results of the wage-regression are presented in Table 6.2, where we consider a version with logarithmic age and education and a version with age and education in calendar years. Given the differences in significance of the age-profiles, we prefer the logarithmic specification, although the difference in climate coefficients is very small. The estimation results for μ and Cantril are presented in Table 6.3.

In the analysis of Table 6.2, education stands for the number of years completed. Comparing the climate block of Table 6.2 with that of Table 6.3, we see that the effect of climate on wages is roughly the same as the effect of climate on μ, suggesting that people are indeed already partially compensated for climate hardship. The smaller number of respondents included in the wage-regressions is due to a number of cases where education is missing.

We can be brief about the non-climate coefficients: regarding the μ-equation these results replicate the findings for other countries bar the unusually large coefficient for family size, which indicates that children have a larger negative influence on financial welfare in Russia than in most countries. This may reflect the recent increase in the relative cost of children in Russia. The non-climate Cantril coefficients indicate that well-being is positively influenced by income and declines with age until 40, after which it increases again.

Table 6.2 *Results of the Full Climatic Model-Estimation for Household Wages: Expected Wages in Different Climate Conditions*

| | Least squares regressions: | | | |
	Log-household Income (1)[1]		Log-household Income (2)[1]	
Intercept	13.2	(4.3)	7.2	(2.4)
Ln(# adults)	0.68	(28.9)	0.72	(23.2)
Ln(age)		–3.41	(5.4)	
Ln²(age)		0.47	(5.4)	
Ln(education)		0.29	(12.0)	
Age			–0.01	(2.8)
Age²			0.00007	(1.6)
Education			0.04	(11.8)
Dummies				
Rural	–0.21	(7.4)	–0.2	(6.9)
Volga and South Russia	0.01	(0.2)	0.04	(1.0)
Wave2	1.18	(39.7)	1.65	(44.7)
Climate variables				
Ln(JANTEMP)	–1.53	(2.9)	–1.32	(2.4)
Ln(JULTEMP)	4.09	(4.2)	4.98	(3.9)
Ln(TEMPDIF)	–2.62	(5.7)	–2.56	(5.4)
Ln(JANWIND)	5.35	(4.7)	5.54	(4.7)
Ln(HEIGHT)	0.04	(1.3)	0.05	(1.7)
Interaction terms				
Ln(JANTEMP)*ln(JANWIND)	–1.54	(4.8)	–1.61	(4.9)
Ln(PREC)*ln(TEMPAV)	0.002	(0.2)	0.003	(0.2)
R^2	0.577		0.579	
N	4127		4127	

[1]   Absolute t-values in parentheses

The 12 hidden organization dummies include state, rented, stock company, joint venture, private company, self- or family-employed, public firms and farms. The difference in the two specifications is whether or not age and education were entered linearly or logarithmically

Turning to the climate variables, a complicated picture emerges from Table 6.3. If we concentrate on climate and m, we first see that the higher the temperature in January, the lower m and thus the greater financial welfare, as expected. Another plausible relationship is that between JANWIND and JANTEMP, whose combination in the m-equation can be read as:

$$4.07 \times \ln(JANWIND) - 1.12 \times \ln(JANWIND) \times \ln(JANTEMP) =$$

$$\ln(JANWIND) \times \big(4.07 - 1.12 \times \ln(JANTEMP)\big)$$

Table 6.3 *Results of Climatic Model-estimation: The Effect of Climate on the Amount of Income Needed to Reach a Fixed Standard of Welfare and the Effect of Climate on Well-Being (Cantril)*

Ordered Probit Regression	μ^1		Order Cantril^1		Cantril^1	
Intercept	−4.35	(2.0)	17.72	(5.3)	–	
Ln(y)	0.62	(56.6)	0.25	(16.9)	0.14	(6.6)
Ln(fs)	0.17	(8.7)	–		–	
Ln(age)	2.14	(5.1)	−3.92	(6.6)	−2.69	(3.2)
Ln^2(age)	−0.30	(5.4)	0.53	(6.6)		
Dummies					0.36	(3.2)
Rural	−0.08	(4.3)	2.17	(4.7)	1.82	(2.8)
Volga and South Russia	0.13	(4.3)	0.08	(1.8)	0.08	(1.3)
Wave2	0.36	(17.8)	−0.22	(7.9)	−0.17	(4.3)
Climate variables						
Ln(JANTEMP)	−1.30	(3.3)	–		–	
Ln(JULTEMP)	3.84	(5.4)	–		–	
Ln(TEMPDIF)	−2.31	(6.9)	–		–	
Ln(JANWIND	4.07	(4.9)	−5.67	(6.5)	−6.75	(5.6)
Ln(HEIGHT)	0.11	(5.5)	−0.11	(3.4)	−0.08	(1.9)
Ln(RAINDAYS)	–		0.86	(6.1)	0.36	(1.9)
Ln(SUNHOURS)	–		0.84	(4.5)	0.37	(1.4)
Ln(WINTERPREC)	–		−0.50	(4.3)	−0.16	(1.0)
Ln(SUMMERPREC)	–		−0.35	(4.6)	−0.22	(2.1)
Interaction terms						
Ln(JANTEMP)*ln(JANWIND)	−1.12	(4.8)	1.59	(6.7)	1.85	(5.7)
Ln(PREC)*ln(TEMPAV)	0.06	(5.5)	0.21	(4.3)	0.09	(1.3)
Ln(HUMIDITY)*ln(TEMPAV)	–		−1.45	(7.0)	−1.29	(4.4)
Rural*ln(PREC)	–		−0.34	(4.6)	−0.28	(2.7)
R^2	0.731		0.094			
McFaddens pseudo-R^2		0.007				
% of observations correctly predicted		55.8%				
N	4412		4412		4412	

^{1} absolute t-values in parentheses
   y denotes current household income not corrected for climate
   For the total effect of climate on welfare and well-being, $\beta_1\theta'AC_i$ must be added to i and $v_i\theta'AC_i$ to Cantril
   Age stands for the age of the respondent
   The percentage of observations correctly predicted denotes the percentage of Cantril observations which were correctly predicted by the ordered-probit model. It gives another measure for the success of the model as the pseudo-R^2, which is rather low.

This implies that the negative effect of strong winds in January on financial welfare increases when January temperatures decrease, an effect we have called the 'chill-factor'. As for the effect of July temperatures and precipitation on μ, the presence of interaction variables and TEMPDIF which depend on the July temperature make it hard to interpret one without the others. To give an

Table 6.4 *The Effects of Changes in Temperature and Precipitation on Welfare and Well-being*

Variable	Welfare constant δ income (%)	Well-being constant δ income (%)
Δ TEMPAV	3.0	−15.6
Δ JULTEMP	8.2	−20.7
Δ PREC	−0.1	−1.0

This table calculates the average of the derivative for each individual in the data set of the full climate model, which incorporates a climate correction for income. Changes in temperatures are measured in Celsius whereas changes in precipitation are measured in millimetres a year

insight into the effect of these interacting variables, Table 6.4 shows whether the effect of an increase in temperature or precipitation is equal to an increase or decrease in the income of the respondents. We calculate how much the change of a climate variable is worth in terms of percentage changes to the income of an individual. This is done by holding welfare and well-being constant while allowing the climate variable to change. Thus we find, using equations (2) and (3), the value of a change in climate.

An increase in the variable TEMPAV of one degree Celsius increases welfare on average just as much as an increase of 3 per cent in income would do. However, an increase in the variable TEMPAV with one degree Celsius decreases well-being by the same amount as a decrease of 16 per cent in income would.

We can see that despite the large positive coefficient of JULTEMP in Table 6.4, an increase in the July temperature actually increases household welfare and thus decreases μ. This is because of the effect of the increase in temperature in July on the average temperature, which in turn affects μ in various ways. Tables 6.3 and 6.4 show that, according to the welfare-criterion, welfare is greatest where the temperature in both January and July is high, at low altitude and with little rain.

A different picture emerges when we apply the well-being criterion: a cold and windy winter and high altitude were deemed negative for well-being, just as they were detrimental to financial welfare. Although not relevant to financial welfare, sunshine was evaluated as being positive, whereas precipitation was on average evaluated as being slightly negative. High temperatures coupled with high humidity reduced well-being without influencing welfare. Rural respondents were unhappier if rain increased.

From Tables 6.3 and 6.4 we can see that the relationship between financial welfare and well-being is weak, reflecting the fact that financial welfare is only one of the components of well-being.

CLIMATE EQUIVALENCE SCALES

Using the results presented in Table 6.3, climate equivalence scales can be computed from equation (2) for any climate in Russia which falls within the range of climates present in our data set. Table 6.5 compares a variety of

Table 6.5 *Climate Equivalence Scales in Several Russian Cities: Relative Incomes, the Relative Cost of Living and the Relative Cost of Well-Being*

Variable	Moscow	Gurjew	St. Petersb.	Dudinka	Novosib.	Cholmsk
JANTEMP	−9.9	−10.4	−7.6	−29.5	−19.0	−9.5
JULYTEMP	19.0	25.4	18.4	12.0	18.7	15.6
TEMPAV	4.5	7.8	4.6	−0.6	−0.2	3.1
TEMPDIF	28.9	35.8	26.0	41.5	37.7	25.1
JANWIND	5.0	6.3	3.4	6.7	4.1	6.8
WIND	5.0	5.5	3.6	6.4	3.8	5.7
RAINDAYS	187	83	196	189	197	199
PREC	568	164	559	267	425	777
HUMIDITY	76	67	77	79	75	75
SUNHOURS	1887	2579	1563	1518	2041	1604
HEIGHT	156	23	4	20	162	29
Equivalence scales						
Current incomes	1.0	0.763	1.133	4.157	1.353	0.995
Leyden-equivalent	1.0	0.505	0.988	5.394	1.335	1.041
Well-being-equivalent	1.0	0.849	1.085	2.463	1.069	0.743

Russian climates by means of showing the climate equivalence scales for six different sites in Russia: Moscow in the centre of old Russia, Gurjew on the northern tip of the Caspian Sea, St Petersburg near the Baltic Sea, Dudinka on the Arctic Ocean, Novosibirsk in the southern part of Siberia and Cholmsk in the extreme east of Russia. The equivalence scales are normalized to Moscow.

From Table 6.5 we see that climate has a strong influence both on financial welfare and well-being. In Gurjew, income is about 24 per cent lower than in a Moscow climate. Gurjew respondents only require about half the income needed in Moscow to maintain the same welfare level, which means that respondents in Gurjew are on average better off. The Cantril scale indicates that a 15 per cent lower income would give respondents in Gurjew the same well-being level as Muscovites. This means that, at present, incomes in Gurjew are undercompensated for their climate, if we proceed from a well-being criterion, and overcompensated from a welfare point of view.

Not surprisingly, Dudinka, a very cold and windy area near the Arctic circle, is the most expensive of these six places in which to live. The cheapest place to live according to the Leyden scale was Gurjew, a warm and dry place at low altitude with a relatively large number of sunhours.

Current incomes concur more with what would be expected from the Leyden-scale than from the Cantril scale, suggesting that respondents are compensated for changes in welfare rather than for changes in well-being.

THE EFFECTS OF CLIMATE CHANGE

As a probe into the question of what the costs and benefits of climate change in Russia might be, we computed what would happen if the average temperature rose by one degree while all other climate variables remained unchanged.[11] If temperatures rose by one degree, inhabitants of Moscow would need 13.5 per cent less income to maintain the same welfare level. Households in Moscow can thus expect to benefit from a rise in temperature. When we calculate the effect on each individual in our data set, we find that, in Russia, the average income impact of a one degree rise in temperature would also be a gain in income. The question is whether there would be an income gain if we look at the total effects of climate change. Ideally we would want to know what changes in other climate variables in different regions would accompany a change in temperature. This would require causal relationships between temperature and other variables which are not yet known (Perman, 1994).[12] The problem is greatest when it comes to computing the effect of climate change on well-being. As Table 6.3 shows, well-being is generated by the interaction of highly correlated variables, whose cause-and-effect structure is unknown. Model-based predictions do exist on the effect of a temperature rise on other climate variables, but the uncertainty of these predictions is too great to be reliable. The climate system is simply too unpredictable at present.[13] We have therefore restricted ourselves to an analysis of the effect of climate change on financial welfare, since we can make reasonable assumptions on the variables involved.

One probable effect of higher temperatures is that they will be accompanied by more precipitation. An often quoted prediction by the Intergovernmental Panel on Climate Change (IPCC) is an increase in temperature of 2.5°C coupled with an increase in precipitation of 8 per cent (Houghton et al, 1990). As these predictions are still very rough, especially for individual regions, several different possibilities have been evaluated in Table 6.6. The effects of these changes in temperature and precipitation are explored by calculating the likelihood that the average Russian respondent would benefit financially from the climate change.[14]

From Table 6.6 we can see that if the climate coefficients of Table 6.3 do not change if the climate changes, Russian households will probably gain from increases in temperature.

It would be wrong to conclude from this table alone that Russia will benefit from climate change, as the changes in costs that households currently incur from climate conditions are only a part of the costs incurred by climate change. Two effects of climate change on household costs which are not considered in this chapter and which may tilt the balance against climate change are transition costs (the costs of adapting to the new situation) and the costs of indirect and as yet unknown effects, such as rising sea-levels.

Table 6.6 *The Likelihood of Financial Gain (%) from Climate Change at Constant Welfare Levels*

Scenario	Precipitation unchanged	Precipitation increased by 5%	Precipitation increased by 10%
Temperature up 1°C	91	87	78
Temperature up 2°C	99	99	97
Temperature up 3°C	99	99	99

These probabilities were derived by calculating the standard estimates of the expected financial benefit for the average incomes in our data set. The formula is derived in Appendix 2. Estimates are rounded downwards.

CONCLUSIONS

The objective of this study was to examine the effect of climate upon operational measurements of the concepts of financial welfare and well-being using a large Russian survey. Our findings are that climate has marked effects which differ substantially from the effects that the climates have been found to have upon financial welfare in Western Europe. This is probably due to the greater range of climate in Russia. In Russia, financial welfare was found to be negatively related to cold and windy winters. Well-being is also influenced negatively by harsh winters, but benefits from the number of sunhours. The factor of high levels of humidity together with high temperatures was also seen to have a strong negative influence on well-being.

The results were used to calculate the effects of climate change on financial welfare, and it was found, not surprisingly, that, under strict assumptions, Russian households can in general expect to incur lower climate costs when the temperature rises.

Similar equivalence scales can be computed for other countries with greatly varying climates such as the US. However, they might well give very different results from ours as the interaction between climate, cultures, soils, industrial practices and agricultural practices will lead to a distinct climate cost structure in each region.

The method employed in this study only needs a household survey across many areas to calculate the costs of different climates. The same method can also be used to assess the money value of other conditions which affect the financial welfare and well-being of households, such as crime, health and public services.

NOTES

1   This chapter is reproduced with kind permission of Kluwer Academic Publishers. It was originally published under the same title in 1998 in *Climatic Change*, vol 39, pp61–81.
2   See Van Praag (1994) for an extensive discussion.
3   Mostly six levels have been used and sometimes even eight or nine. In this Russian survey it was decided together with CESSI to use only five levels.
4   Previously it was found, and also confirmed for the data set used in this chapter, that σ_j^2 is only weakly dependent on y_{ic}, fs_i, age_i or *education*; we shall therefore use the population

average σ^2 for interpersonal comparisons (Van Praag (1971), Van Praag and Kapteyn (1973), Hagenaars (1986)).

5   Cantril's original question contained 11 vertical levels, from 0 to 10.

6   This method of deriving equivalence scale was developed by Plug and van Praag (1995a) and Van Praag and Plug (1995).

7   Note that differential unemployment rates or levels of development across regions, for example, should not have any effect on the level of financial need, bar the effect that is catered for by the inclusion of the effect of income on financial need.

8   Here we have taken the two waves as independent cases, as it was found that we can accept the hypothesis of equal coefficients in both waves and that the corresponding error terms in both waves were approximately independent. For more information, see Appendix 2.

9   The 1988 paper uses the EUROSTAT 1979 survey commissioned by the European Commission, which was extensively discussed by Hagenaars (1986).

10 Unfortunately, the 1979 data set is not available any more for further analysis.

11 Using equation (3), we could explicitly calculate the monetary benefits of climate changes for any individual or region in the data set. Given that we only had 35 different climate regions to work with, however, we only look at whether we can expect a gain or a loss.

12 Some relationships are known. For example, high altitudes influence temperature negatively, but temperature does not influence altitude. The relationship between wind and temperature is not so clear-cut.

13 We would like to thank David Maddison and Richard Tol for pointing out the existence and limitations of the existing predictions.

14 The assumption we make about the causal structure of climate is that in January wind remains unaffected by changes in temperature and precipitation. A change in average temperature is assumed to be a consistent change in the temperature throughout the year.

Chapter 7

A HEDONIC STUDY OF THE NON-MARKET IMPACTS OF GLOBAL WARMING IN THE US

By Robert Mendelsohn

During recent years, a second generation of impact studies has re-examined the effect that climate change would have on the American economy.[1] In contrast to the first wave of impact studies (see Smith and Tirpak, 1989), these studies suggest that modest warming would provide small benefits to the American economy. There are four reasons for these more optimistic results. First, the studies incorporate efficient adaptation that reduces damages and increases benefits. Second, they are more comprehensive and include a larger set of benefits as well as damages. For example, 'warm-loving' vegetables and fruit were included in the new agricultural studies along with the 'cool-loving' grains studied in earlier analyses. Third, the new studies use dynamic models to examine capital intensive sectors such as forestry and coastal structures instead of the earlier comparative static models. Fourth, the science underlying climate impacts has changed, indicating that at least vegetation will prosper in a mildly warmer, wetter, CO_2 enriched world.

To date, however, the second generation studies have largely been limited to studying market impacts in the US.[2] Climate change, however, is expected to affect more than just market phenomena. Warmer climates will alter ecosystem boundaries, change local weather, and possibly affect health as well. Preliminary economic analyses of some of these quality-of-life effects suggest large damages. For example, the IPCC (1996b) estimates that mortality from heat stress would result in from US$6 to 37 billion of damages per year in the US alone. This same report would add up to US$12 billion of additional damages from less pleasant weather (human amenity) and another US$0.5 to 1 billion of migration costs as people flee warmer climates each year.

This study tests the hypothesis that warmer climates would reduce American non-market values. A hedonic analysis of American wages and housing prices is conducted.[3] The study adopts the standard assumption in this literature that there is a national competitive housing and labour market. In order for firms to attract labourers to their area, they must bid against firms in other locations. Labourers then consider the entire package of amenities, housing prices, and wages associated with working and living in an area. To remain competitive, locations with fewer amenities must offer higher wages/ lower housing costs to have a competitive package. Housing prices and wages adjust to achieve an equilibrium (see Rosen, 1974, and Roback, 1982). Hedonic analyses of these equilibrium rents and wages reveal the marginal value that local people place on different attributes of living where they have chosen.

Past studies of American wages have revealed that people prefer warmer climates.[4] However, climate was incidental to most studies (except for Hoch and Drake, 1974), so the result has not been considered definitive. Further, comparisons across studies are difficult to make because each study used a different measure of climate. In this study, we use hedonic wage and housing models to explore the relationship between climate, rents and wage rates. Instead of studying a limited number of cities, we examine over 3000 counties in the continental US. By extending well beyond metropolitan areas, we get significantly more variation in climates and more chances to disentangle social attributes from climate. This comes at the expense of relying upon aggregate (county-wide) rather than individual data. However, given that we are trying to determine which county people select to live in, not which job and house they choose, the aggregate data may not entail too great a sacrifice. We also explore a more complete description of temperature and precipitation than earlier studies. Using an extensive data set based on US Census information, we regress wage rates on climate and other control variables. The results are then used to quantify the non-market impacts of several climate change scenarios for the US.

Although this study captures many non-market impacts associated with choosing one climate versus another, it has several limitations. First, the study focuses on values associated with each person's locality. Impacts which occur outside the US and impacts which are generic (not location specific), such as the loss of a species, are not measured. Second, the study only reflects aspects of climate that people are aware of. Thus, it clearly includes weather, local ecosystem type, and the possibility of heat waves, but does not include indirect impacts that people are not aware of, such as vector-borne disease risks. Third, the scope of the impacts captured in this study includes effects which have been captured in analyses of recreation and residential energy demand, so that care must be taken not to double count these effects. Finally, the methodology measures marginal values for job and housing attributes including climate. Because the underlying structural equations are not revealed, the hedonic price approach will under- or overestimate the precise value of non-marginal reductions or increases in goods and services (see Rosen, 1974).

The next section of the chapter briefly reviews the theory underlying the hedonic wage model. The third section then presents the empirical methodology and results and calculates the aggregate non-market impacts associated with different climate scenarios. The concluding section discusses the policy implications of the results.

Theory

The hedonic wage model (Rosen, 1974) assumes a competitive national labour market. Firms in each location and sector take wages as given and determine the number of jobs they can offer. The aggregate demand for labour, L_i, the sum of firm demand functions, in each location, i, is a function of wages, W, and other attributes associated with the location, Z:

$$L_i = D(W_i, Z_i) \qquad\qquad 1$$

The derivative of (1) with respect to wages is expected to be negative. Although firms are not constantly choosing to relocate, we assume that marginal firms can expand or contract activities in different locations. Thus, the different locations do compete with one another for labourers.

We assume that labourers examine their alternatives and select a sector of the economy most suited to their tastes and abilities. Given the sector of their choice, labourers consider working and living in different locations given the wages being offered. We assume that workers consider all the important features of a location when making this choice, including not only the characteristics of the job being considered but also the salient features of living in that location. Aggregating across workers, one gets a supply of labourers at each location that depends on wages and the attributes of living and working in that location, X, relative to all others:

$$L_i = S(W, P, X) \qqquad\qquad 2$$

Although not everyone considers living in every location across the country, it is sufficient for our purposes that a number of marginal workers consider moving at any time.

Housing is made available to people as residential uses compete with alternative land uses. If more people want to live in a city, the city will have to expand, increasing land scarcity. This results in a rent gradient, which increases as one approaches the centre of the city. The larger the city, the higher the rent gradient, all other factors held constant. Because it is necessary to reside in proximity to one's work, housing prices play a role in locational decisions. Although the precise rent people will pay in a city depends upon where they choose to live relative to the centre, higher rent gradients lead to higher average rents, P.

The individual household must solve a wage and housing decision simultaneously when choosing what county to live in. Aggregating across all households, the solution is an equilibrium set of wages and rents for each location. For a set of wages to be in equilibrium, the marginal firms must be content with remaining at the location of their current operations and the marginal workers must be content to live and work where they have chosen. At any location, the wage and rental rates must compensate for any perceived differences in amenities. Examining the broad set of job packages offered across the country, one can observe the marginal value of individual amenities by examining the equilibrium rents and wages across locations in relation to the set of attributes offered:

$$W_i = H(P_i, X_i) + u_1$$
$$P_i = R(W_i, X_i) + u_2$$

3

In this analysis, we specifically wish to explore the relationship between rents and wages and climate. We consequently include a detailed account of climate, including seasonal temperatures and precipitation, for example, in the hedonic wage and rent functions. These more detailed measures not only capture what citizens prefer in direct weather but also climate-related ecosystem effects. Our underlying hypothesis is that areas with better climate have a combination of lower wages and higher rents – climate is a fringe benefit of living in such locations. For example, if people prefer warmer temperatures, warmer places will have lower wages but higher rents. Because we expect that climate has a curvilinear relationship with wages, we introduce a quadratic function describing climate variables. For example, we expect that people who live in cool places would prefer a warmer summer. However, it is also likely that people who already have a hot summer would not want it much hotter. The quadratic function can capture this complex response function.

METHODS

In order to estimate the hedonic wage and rent model for a wide variety of climate conditions, we turn to county data across the continental US for this analysis. The county data offer the opportunity for almost 3000 observations in a wide variety of climate settings. In contrast, most earlier hedonic studies focused on SMSAs that had individual wage and price information. Although these urban data sets provided more micro-information about job characteristics, only a limited number of cities were available, which restricts the opportunity to distinguish subtle climate effects.[5] Further, by using counties, we can explore a wide range of temperature and precipitation values. For example, annual temperature has a mean of 12.6°C in the US, but the counties in the sample range from 2.8° to 24.9°C. Climate warming scenarios of 1.0–3.5°C thus lie well within the range of the sample.

In an attempt to try to limit differences across job types, we examine four sectors independently. At the county level, census data collected in the *City and County Data Book* distinguish between 1987 wages paid to retail, wholesale, service, and manufacturing jobs. Separate analyses were conducted for each sector. In addition, one equation measures rents.

In order to explain the different wages and rents paid across counties, a wide range of control variables is explored which capture characteristics of the work force such as the percentage of female workers, education levels, characteristics of the location such as the percentage of single family homes, the crime rate and altitude, and measures of cost of living such as purchased water and the percentage urban area for each county. A complete list of the variables included in the model, their means, and their definitions are given in Table A3.1 in Appendix 3. Care must be taken in interpreting the coefficients of the control variables in the data set because many may reflect more than one influence. For example, the percentage of families headed by a female could reflect a characteristic of the work force, family structure, or some other unmeasured attribute. Because we do not believe these variables are being affected by climate, we leave them in the regression to remove unwanted variation.

In addition to the control variables, we include a complete set of seasonal temperature and precipitation variables. Each climate variable reflects the 30-year average weather associated with a month. For example, January temperature is the average January temperature in °F observed from 1950–80. These values were fitted to each county from thousands of weather stations using a sophisticated interpolation scheme (see Mendelsohn et al, 1994). Each month was selected to capture the influence of a season (winter, spring, summer, and fall). A squared term was also introduced to allow the climate response function to be non-linear. All the climate variables have been demeaned to facilitate interpretation of the coefficients. The marginal influence of a one degree increase in April temperature above the US mean is thus captured by the linear coefficient on April temperature. The squared terms capture whether the response function is U-shaped (positive) or hill-shaped (negative). The complete functional form for the hedonic functions is:

$$\log(W) = a_0 + \sum_{i=1}^{8} a_i c_i + \sum_{i=9}^{16} a_i c_i^2 + \sum_{j=1}^{14} a_j c_j + \sum_k P_k \qquad 4$$

where c_i are seasonal climate variables, x_j are control variables, and P_k represents the prices of the other hedonic equation as stated in (3). Because the prices, P_k, are endogenous, we estimate the equations with two stage least squares.

Although there are 3111 counties in the continental US, the samples in the regressions are smaller because some counties are so sparsely populated that few employment opportunities exist. We weight each observation in the wage regressions by the number of employed people in that sector and in the rental regression by the number of households. Some counties, however, had no employment in certain sectors. This limitation mostly affects the manufacturing regressions, almost 800 counties having little or no manufacturing activity. Rather than limit the analysis to counties that had all four types of employment, we analysed each equation separately.

The optimal specification for a model of this complexity is not clear. Locational choice is dictated by many considerations. In order to explore whether the results are robust across specifications, we present three alternative models. The first model includes the basic set of climate and control variables. The second model adds a set of regional dummies.[6] These dummies control for some alternative spatial hypotheses such as the lower incidence of unions in the southeast. The third model introduces climate variation measures. These measures include the diurnal cycle, the interannual range of precipitation, and the interannual range of temperature. This third model tests the sensitivity of non-market values to climate variation.

Results

The results of the basic regressions are displayed in Table 7.1. The coefficients of the control variables generally have their expected sign. For example, higher education is generally associated with higher wages, presumably because a more skilled labour force is more productive. Higher female labour force participation rates resulted in lower wages, reflecting discrimination, a larger work force, or some unmeasured attribute correlated with working women. Urban settings and counties with larger populations have higher wages that could reflect disamenities from bigger cities. Higher percentages of single family homes and older homes were generally associated with lower wages, presumably because of higher amenities. Private water delivery leads to higher wages, possibly because of the higher cost compared to wells. Counties that were growing and had many recent arrivals generally have lower wages, reflecting either an unmeasured amenity drawing people to an area or a temporary disequilibrium. Finally, wages are lower in higher altitudes, reflecting either lower costs of living or amenities associated with living in higher locations.

Theory does not specify whether the factors that make a place attractive would encourage both lower wages and higher rents. Theory merely implies that the net effect of attractive features should be balanced by a combination of differential rents and wages. The coefficients on the housing equation are the opposite sign of the coefficients on the wage equation with only two control variables: female participation rates and female head of household. For the other control variables, the housing coefficient had the same sign as the wage coefficient. In general, factors that encouraged higher wages also encouraged higher housing prices. It is possible that several of these variables are proxies for cost of living that might affect both housing prices and wages in the same direction. It is also possible that some variables are acting as proxies for different omitted variables in each equation. For example, single family homes may be acting as a housing attribute in the housing equation, but as a proxy for suburban employment in the wage equation.

The focus of this analysis, however, is upon climate. As can be seen in Table 7.1, many of the temperature and precipitation variables are significantly different from zero. The effects vary by season and they often vary across equations. Places with warmer January temperatures generally had higher wages, but places with warmer Aprils had lower wages. Places with warmer temperatures in October have higher wages. People apparently like to have a warmer spring but a colder winter and fall compared to the US mean. The results in summer are mixed. One must be cautious in interpreting these results because they reflect both what climate people prefer as well as the resulting ecosystems. For example, a cold winter reduces insect populations and alters which plant species will survive. If people choose a location because it has these characteristics, they are voting for a cold winter even if they do not know about this connection.

In order to test the robustness of these results, we also explore two other regression specifications. The first alternative (see Table A3.2 in Appendix 3) includes a set of regional dummies to control for regional influences such as the degree of unionization. The south-central region is omitted. The second

Table 7.1 *Basic Hedonic Regressions*

	Rents	Service wages	Retail wages	Wholesale wages	Manufacturing wages
JAN TEMP	22.9 (7.21)	1.80 (0.45)	4.60 (2.19)	6.62 (2.00)	−9.65 (1.87)
JAN T SQ	0.30 (2.29)	−0.36 (2.20)	−0.19 (2.22)	0.13 (0.96)	−1.52 (6.96)
APR TEMP	−115. (33.99)	−13.2 (3.21)	−36.4 (16.28)	−25.8 (7.11)	−3.43 (0.62)
APR T SQ	−1.58 (4.42)	−0.55 (1.20)	−0.84 (3.46)	−1.92 (4.63)	−2.04 (3.38)
JUL TEMP	5.44 (1.57)	−3.64 (0.89)	3.33 (1.45)	7.67 (2.11)	−27.5 (4.64)
JUL T SQ	−1.12 (4.41)	−0.23 (0.79)	0.35 (2.08)	0.68 (2.53)	0.08 (0.18)
OCT TEMP	80.0 (12.73)	16.0 (2.10)	29.0 (6.91)	2.19 (0.33)	22.6 (2.20)
OCT T SQ	−0.73 (1.42)	0.23 (0.35)	0.46 (1.34)	1.28 (2.16)	5.79 (6.59)
JAN PREC	−0.04 (0.21)	−0.21 (1.05)	0.32 (2.74)	0.04 (0.21)	0.23 (0.81)
JAN P SQ	−.006 (8.40)	−.002 (2.37)	−.004 (8.57)	−.003 (3.40)	.000 (0.17)
APR PREC	−0.41 (2.26)	−0.44 (1.99)	−0.09 (0.78)	0.35 (1.85)	−0.19 (0.63)
APR P SQ	.017 (6.54)	.002 (0.72)	.008 (4.69)	−.007 (2.51)	−.022 (4.95)
JUL PREC	−0.03 (0.32)	−0.11 (1.00)	−0.44 (7.24)	0.15 (1.53)	0.89 (5.76)
JUL P SQ	−.001 (1.06)	.005 (3.14)	.004 (4.79)	−.006 (3.83)	−.014 (5.56)
OCT PREC	1.90 (9.17)	0.87 (3.28)	1.31 (9.46)	0.12 (0.51)	−2.41 (7.19)
OCT P SQ	−.003 (1.16)	.002 (0.69)	−.001 (0.66)	.005 (1.79)	.017 (3.92)
CONSTANT	7830. (111.39)	10000. (104.27)	9420. (188.06)	10300. (120.83)	10400. (80.65)
HIGH SCHL	5.78 (10.38)	2.04 (2.59)	−0.60 (1.51)	6.41 (9.67)	15.1 (17.21)

Table 7.1 *(Continued)*

	Rents	Service wages	Retail wages	Wholesale wages	Manufacturing wages
COLLEGE	10.3	11.7	5.29	9.85	3.36
	(21.67)	(20.88)	(16.48)	(19.01)	(4.38)
ALTITUDE	−5.98	−1.63	−0.92	−3.30	−4.63
	(16.19)	(3.74)	(3.79)	(8.14)	(7.55)
FEMALE PART	4.50	−16.6	−0.90	−23.3	−42.7
	(2.88)	(7.61)	(0.81)	(12.34)	(15.36)
URBAN	79.2	115.	31.1	13.2	86.3
	(12.07)	(11.96)	(6.81)	(16.21)	(9.01)
SINGLE HOMES	−2.34	−1.46	−1.36	−1.15	1.89
	(9.97)	(4.78)	(8.15)	(4.42)	(4.90)
OLD HOMES	−3.89	−1.50	−2.63	−2.76	−3.14
	(13.66)	(4.09)	(13.43)	(8.79)	(7.30)
WATER CO	1.05	1.84	−0.17	1.93	2.51
	(6.29)	(7.48)	(1.37)	(9.42)	(10.08)
POP CHANGE	2.18	−0.09	0.73	−0.32	−1.70
	(13.68)	(0.43)	(6.46)	(1.74)	(5.69)
RECENT MOVE	−3.25	−4.66	−4.38	−2.87	−1.43
	(6.83)	(7.91)	(13.71)	(5.54)	(1.97)
POP	3.32	26.3	5.84	9.89	2.08
	(1.88)	(14.01)	(5.04)	(6.13)	(0.85)
FEM HEAD	−3.28	5.76	2.00	6.87	14.8
	(6.57)	(9.06)	(5.74)	(12.50)	(19.51)
FAMILIES	3.26	−1.59	−0.02	0.51	−0.36
	(6.04)	(2.27)	(0.05)	(0.76)	(0.36)
R^2	0.849	0.817	0.781	0.842	0.766
N	2179	3019	3095	2754	2358

The dependent variable is the log of average annual 1987 wages or rents
The coefficients have been multiplied by 1000 for ease of presentation
T-statistics are in parenthesis.

alternative (see Table A3.3 in Appendix 3) adds a number of additional climate variables including the interannual variation of precipitation and temperature and the diurnal range of temperature in each season. Table 7.2 compares the marginal climate coefficients in all three models evaluated at the US mean.

The temperature effects on wages are slightly more positive in the regional compared to the basic model because the interregional correlation between temperature and wages is negative. The remaining relationship between temperature

Table 7.2 *The Annual Marginal Impact of Climate*

Variable	Sector	Temperature			Precipitation		
		Basic	Regional	Variance	Basic	Regional
RENTS	−34.6	14.6	−110.	7.7	3.1	−10.0
	(8.6)	(3.1)	(9.8)	(1.1)	(1.2)	(1.5)
SERVICE	19.1	87.1	−9.1	1.9	5.4	−1.5
WAGES	(37.2)	(42.2)	(41.5)	(4.7)	(5.3)	(6.2)
RETAIL	4.9	26.8	−29.6	11.0	12.6	−2.7
WAGES	(10.8)	(12.1)	(12.5)	(1.3)	(1.5)	(1.9)
WHOLESALE	−253	−116	−434	17.6	21.9	−3.8
WAGES	(47.4)	(53.7)	(55.7)	(5.8)	(6.6)	(8.9)
MANUFACTURING	−605	−506	−987	−49.8	−43.9	−65.1
WAGES	(86.7)	(100.)	(96.8)	(10.6)	(12.2)	(14.9)

The annual marginal impact is the sum of the seasonal linear coefficients times wages (per °C or mm/month) evaluated at the US mean

A positive (negative) value in the wage equations and a negative (positive) value in the rent equation imply a harmful (beneficial) effect

The standard error is in parenthesis.

and wages, even controlling for regions, remains negative. Precipitation is hardly affected by including regional variables. The marginal temperature and precipitation effects have become more negative in the climate variance model because the climate variation variables are correlated with the mean temperature and precipitation variables. Reviewing all three regressions, warmer temperatures are generally associated with lower wages and lower rents. Temperature effects are larger than precipitation effects. The wage effects are generally larger than the rental effects. The standard error of each estimate is presented in parenthesis, revealing that most of the marginal effects are significant except for the service sector.

Examining the coefficients of the squared climate terms in the regressions (Table 7.1) reveals that they are significantly different from zero. The signs of the second order terms, however, vary across sectors with the service, retail, and wholesale sectors having a predominance of negative squared terms and only the manufacturing sector having net positive squared terms. The service, retail, and wholesale sectors thus exhibit a concave shape whereas the manufacturing sector has the expected convex response function. The rental equation has an expected concave shape. These quadratic terms become more important with more severe warming scenarios, reducing marginal benefits overall.

In order to get some sense of the magnitude of climate effects, we explore six climate scenarios reflecting the range of outcomes predicted by climate scientists (IPCC, 1996a). The scientists predict a best guess increase of 2°C by 2100. They also provide a low estimate of 1°C and a high estimate of 3.5°C. With these three temperature alternatives, we couple 8 per cent and 15 per cent increases in precipitation for six total scenarios. We assume the climate change is uniform across the US and across seasons. Although a perfectly uniform change is not likely, it is a close approximation to the expected prediction across a large set of climate models (see Williams et al, 1997). Of course, aggregate impacts would be sensitive to alternative scenarios. For example, more warming in northern areas would result in higher benefits and relatively more warming in southern areas would produce more damages.

The results of the six climate scenarios are presented in Table 7.3 using each of the three regression models. The individual sector results have been adjusted to account for the transformation of the dependent variable.[7] The aggregate estimate for each wage sector is made by weighting the impact in each county by the employment in that sector. The aggregate estimate for housing weights counties by households. The individual sector wage estimates are summed and adjusted for all employment in the US (the measured sectors account for half of all employment). The rental effects are added to the wage effects for a total net welfare estimate. Table 7.3 reports the 95 per cent confidence interval for each estimate. On the low end, there could be damages in every sector except manufacturing. On the high end, every sector except retail would enjoy benefits from warming.

Table 7.3 *The Range of Annual US Non-Market Impacts from Global Warming (US$ billion, 1987)*

Scenario		Rents	Service wages	Retail wages	Wholesale wages	Manufacturing wages	Total
1.0° 8% P	Low	−7.5	−3.8	−2.3	−0.6	3.7	2.1
	High	4.0	3.8	0.3	2.4	18.9	39.5
2.0° 8% P	Low	−14.9	−2.8	−2.2	0.3	7.8	6.5
	High	−3.1	4.8	0.5	3.3	23.3	44.6
3.5° 8% P	Low	−32.0	−1.6	−2.3	1.1	7.8	−3.0
	High	−17.4	8.3	1.3	5.0	28.0	45.9
1.0° 15%P	Low	−6.8	−5.5	−3.9	−1.1	6.3	6.1
	High	8.6	4.5	−0.3	2.9	26.5	55.6
2.0° 15% P	Low	−14.6	−4.9	−3.9	−0.3	9.7	9.0
	High	1.8	5.8	−0.0	3.9	31.4	62.0
3.5° 15% P	Low	−32.0	−3.7	−4.0	0.5	9.7	−1.0
	High	−12.5	9.3	0.8	5.7	36.1	63.5
1.0° 8% P	Low	5.8	−4.7	−2.6	−1.2	3.2	−0.9
	High	−3.7	1.6	−0.3	1.4	15.9	30.3
2.0° 8% P	Low	3.9	−5.5	−3.1	−1.1	6.4	1.1
	High	−6.5	1.3	−0.6	1.7	20.1	34.9
3.5° 8% P	Low	−2.4	−6.7	−4.3	−1.3	6.4	−9.4
	High	−15.6	2.1	−1.0	2.3	24.6	34.8
1.0° 15% P	Low	8.8	−6.3	−3.8	−1.6	5.6	1.4
	High	−3.6	1.7	−0.8	1.7	21.7	41.1

Table 7.3 *(Continued)*

Scenario			Rents	Service wages	Retail wages	Wholesale wages	Manufacturing wages	Total
2.0° 15% P	Low		7.2	−7.4	−4.5	−1.6	8.1	1.8
	High		−6.8	1.7	−1.1	2.1	26.5	47.0
3.5° 15% P	Low		0.7	−8.4	−5.6	−1.8	8.6	−7.8
	High		−15.9	2.3	−1.5	2.7	30.5	45.8
1.0° 8% P	Low		−20.1	−3.1	−0.9	0.3	8.8	6.0
	High		−7.4	4.5	2.1	3.4	25.3	46.0
2.0° 8% P	Low		−31.4	−1.9	−0.6	1.6	17.1	17.0
	High		−18.0	5.7	2.6	4.9	35.4	60.5
3.5° 8% P	Low		−51.9	−0.7	−0.4	3.0	25.3	22.4
	High		−35.1	9.7	3.6	7.4	47.4	76.3
1.0° 15% P	Low		−25.2	−4.4	−1.3	0.2	12.3	8.9
	High		−9.2	5.6	2.6	4.1	33.2	59.8
2.0° 15% P	Low		−36.7	−3.5	−1.1	1.4	19.9	18.6
	High		−19.4	7.0	3.2	5.8	43.6	75.3
3.5° 15% P	Low		−57.1	−2.3	−1.0	2.9	28.1	24.1
	High		−36.4	11.0	4.3	8.2	55.3	90.7

Range is 95 per cent confidence interval conditional on climate change

Total assumes independence across sectors and that the measured sectors (50 per cent of all workers and 90 per cent of all households) are representative of unmeasured people

Negative numbers imply reduction in welfare

The aggregate effect reported in Table 7.3 (adding wages and rental figures together) reveals that warming is likely to produce benefits in every climate outcome except a 3.5°C warming. Even in this more severe scenario, the chance of non-market damages from warming is actually quite small. The results across three different specifications suggest that warming is likely to provide non-market net benefits for US citizens. However, the effects are small relative to gross natural product (GNP) averaging about US$25 billion or 0.5 per cent of GNP.

The climate variation regressions also provide a measure of the importance of climatic variation. Global warming is expected to decrease the diurnal cycle by warming up nights more than days. It is unknown what effect climate change will have on interannual variation. Some climate models predict that variation will increase while others predict it will decline (IPCC, 1996a). The marginal value and standard errors of an increase in variation are presented in Table 7.4. An increase in precipitation variance would raise wages slightly, suggesting a small net damage. An increase in interannual or diurnal temperature variation has mixed effects and is generally insignificant despite the significant individual coefficients. The coefficients in each season have opposite signs and so cancel each other out. It is consequently unclear what effect a change in climate variation would have.

Table 7.4 *The Marginal Value of Increased Variance*

Sector	Interannual Precipitation	Interannual Temperature	Diurnal Temperature
RENTS	2.0 (0.5)	−124.0 (17.2)	−20.2 (11.1)
SERVICE	3.7 (1.2)	70.0 (91.2)	−26.4 (41.7)
RETAIL	2.1 (0.6)	−108.0 (24.8)	−47.6 (12.3)
WHOLESALE	2.3 (3.4)	−134.0 (143.)	15.3 (66.9)
MANUFACTURING	10.3 (4.6)	218.0 (288.0)	0.0 0.0

Marginal values of increased year-to-year variation evaluated at the US mean
A positive (negative) value in the wage equations and a negative (positive) value in the rent equation imply a harmful (beneficial) effect
Standard errors are in parenthesis

CONCLUSION

Past hedonic wage studies of urban areas implied that wages were slightly lower in warmer places, implying an American preference for warming. This analysis extends this research by carefully examining the impact of climate on wages and rents in almost every county in the contiguous US. The analysis confirms these earlier studies, showing that warmer and wetter places generally have net lower wages and housing prices. These results imply that Americans prefer warmer and wetter locations. These preferences reflect both a choice about weather as well as the resulting ecosystem and health conditions.

Care must be taken not to double count these results with other non-market and market studies reflecting household choice of location. For example, it is likely that the wage package includes the effects of recreation, residential energy, and known health effects such as heat stress. Mild warming is expected to have harmful effects on residential energy (Morrison and Mendelsohn, 1999) and heat stress (IPCC, 1996b) but beneficial impacts on recreation (Mendelsohn and Markowski, 1999). These net impacts are embedded in the wage study results.

The results indicate that the hedonic price function is quadratic in temperature. This implies that the marginal value of higher uniform temperatures would not itself be uniform. Places that are already quite hot would actually drop in value. Places that are currently quite cold would increase in value by more than average. The curvature of the response function also suggests that more severe warming scenarios would result in damages, even if mild scenarios lead to benefits. The hedonic equation, however, is only a marginal measurement, so that welfare measures for large changes in temperature are likely to be biased. This bias is likely to be small for the climate scenarios considered here, but might not be small for more severe scenarios.

Applying these results to a set of realistic climate scenarios reveals that 2°C warming will result in annual net non-market benefits in the US of between US$1 billion and 75 billion. Given the range of predicted temperatures (IPCC, 1996a), warming will probably provide non-market benefits to Americans for the next century. Combined with the results of recent market studies indicating net benefits to market sectors, the analysis suggests that the modest warming of the next century from greenhouse gases is likely to be beneficial to the US.

Whether there are sufficient global damages from greenhouse gas emissions to warrant an expensive abatement strategy is becoming increasingly questionable. Recent research indicates that temperate developed countries are likely to benefit from warming over the next century. Attention must now focus on measuring the impacts of climate change on subtropical and tropical developing countries. If the damages to developing countries are small, the optimal global strategy for greenhouse warming for the next century may not entail substantive abatement. Depending upon the results of studies in developing countries, however, compensation and adaptation programmes may be a more efficient and equitable alternative.

NOTES

1   Mendelsohn and Neumann (1999) contains a comprehensive set of second-generation market studies of the impacts of climate change on agriculture, energy, forestry, recreation, water, and sea level rise effects in the US.
2   A notable exception is the analysis of recreation (see, eg, Mendelsohn and Neumann, 1999).
3   See Hoch and Drake, 1974, Cropper, 1981, Roback, 1982, 1988, and Blomquist et al, 1988.
4   Roback (1982), Cropper and Arriaga-Salinas (1980), Cropper (1981) and Hoch and Drake (1974) have all found that warmer climates reduce wages, implying people would prefer marginal warming.
5   Roback (1982), for example, analysed 98 cities.
6   I wish to thank Professor Weitzman for this suggestion.
7   The exponent of the predicted log of wages is multiplied by exp (SSE/n/2) to adjust for the transformation of the dependent variable.

Chapter 8

THE AMENITY VALUE OF THE CLIMATE OF INDIA: A HOUSEHOLD PRODUCTION FUNCTION APPROACH

The suspicion is that those most vulnerable to the impacts of climate change are living in subtropical and tropical developing countries. In such countries the assumptions required to retrieve amenity values from land prices and wage rates (perfect information about the characteristics of alternative locations, well functioning markets for land and labour and the absence of barriers to mobility, for example) are unlikely to be met. In such instances the HPF approach, in which consumption patterns are analysed to yield information on the amenity values, seems more applicable. India in particular seems a good test bed for the HPF approach, partly because it includes climatically diverse regions and partly because it has detailed regional information on household expenditures using consistently defined consumption categories.

This chapter uses a cross section of consumption data from the different states and territories of India and combines them with climatological data to determine the amenity value of climate in India. Combining these amenity values with predictions regarding climate change, the change in the cost of living implied by various scenarios is calculated.

EXISTING LITERATURE

Previous attempts to employ the HPF methodology to value climate amenities can be found in Shapiro and Smith (1981) and in Maddison (1997). Both papers involve placing restrictions on the utility function necessary to be able to measure changes in amenity values. Bradford and Hildebrand (1977) demonstrate that provided there exists a price vector, such that marginal changes in the level of the amenity do not affect the level of utility, then amenity values can be recovered from knowledge of market demands. Bradford and Hildebrand refer to this condition as 'demand-dependency'.

The paper by Shapiro and Smith (*op cit*) is best regarded as a methodological proposal rather than a serious attempt to empirically determine amenity values for climate variables. They confine their analysis to different regions within California where the differences in temperature and precipitation are insufficient to identify differences in the cost of living associated with these variables.

The paper by Maddison (*op cit*) began by examining differences in consumption patterns between 60 countries such that major differences in climate were observed in the data set. He found a significant role for climate variables and his analysis pointed unequivocally to an 'optimal' climate of between 15 and 20°C for annually averaged mean daily temperature. Departures from this optimal climate result in higher costs of living, particularly with regard to energy costs. He discovered that the costs imposed by a climate which is too hot or too cold depend upon income levels rather than in the nature of fixed costs. Higher income countries would be willing to pay more to reach the climatic optimum.

Unfortunately, the research presented a number of shortcomings which cast doubt on the results. First of all, the paper by Maddison dealt with average consumption patterns in entire countries. To conduct his analysis he found it necessary to determine the population-weighted climate variables given that most of the countries contained different climatic regions. He achieved this by taking the climate records for each city of over one million people and finding a population-weighted average of the climate of these cities as representative climate of the country. How well this procedure worked is not clear, but in certain countries (eg India) the majority of the population still lives in rural areas. The second issue was that many of the countries in his data set have highly uneven distributions of income. He assumes that the consumption patterns of these countries were generated by a 'representative consumer'. The third issue arises because of the LES that he chose as a vehicle for his work (Stone, 1954). Since the majority of the world's poorest countries are found in the hottest regions, climate variables might have been compensating for the deficiencies of the functional form.

This chapter is an attempt to correct some of the deficiencies of earlier work. In particular, the data for this empirical exercise involve much smaller geographical regions, thereby avoiding the need to engage in any kind of weighting procedures. The climate variables are therefore a much better representation of the actual climate experienced by people living in a region. The need to make assumptions regarding the existence of representative consumers instead of making use of average expenditures only within narrowly defined income deciles is also avoided. Having a relatively large number of observations makes it possible to fit an extremely flexible functional form to the data in which Engel curves are allowed to take a variety of shapes. At the same time alternative ways of representing a fluctuating amenity such as climate are explored. There is one shortcoming however: the data set contains no variation in prices and is purely cross-sectional in nature. This does not preclude estimation of the costs of living associated with changes in the level of environmental amenities but does impose certain restrictions.

Extending Systems of Demand Equations to Reflect the Role of Environmental Amenities

The procedure used to incorporate environmental variables into systems of demand equations used by Maddison (1997) is borrowed directly from the literature on the incorporation of demographic variables into systems of demand equations. More specifically, the analysis uses 'demographic scaling' (see for example Pollak and Wales, 1981). This is not the only way of including environmental variables in demand functions, but the advantage of this approach is that the nature of the role played by the use of established utility functions, whose limitations can constrain the results in important ways, is already well understood. Scaling replaces the original utility function u with:

$$u = u(q_1 / d_1, q_2 / d_2, \ldots)$$

where u is utility, q_i is the quantity of good i and the d's are the scaling functions given by:

$$d_i = d_i(z) = \sum \eta_i z_i$$

in which z is a vector of environmental amenities and η is a vector of parameters. This corresponds to an indirect utility function of the form:

$$u = v(p_1 d_1, p_2 d_2, \ldots y)$$

where p is prices and y is income. It involves replacing the original system of demand equations q^i with:

$$q_i / d_i = q^i(p_1 d_1, p_2 d_2, \ldots y)$$

The HPF interpretation of scaling the quantities of marketed commodities in this manner is that instead of having a utility function dealing with consumption of marketed goods, utility is better thought of as dependent upon the flow of services derived from combining marketed goods and environmental amenities in a simple technical relationship. Changes in the cost of these services result in substitution either to or away from other services.

In order to identify the individual scales it is generally necessary to have data in which there is some price variation. But without any price variation it is still possible to determine amenity values provided that the scaling functions d_i are the same across all commodities. This is not very realistic, but may serve as an approximation. Moreover, if the definition of commodity aggregates changes over time or such surveys are conducted only at infrequent intervals (as is the case in India) then this is the only possible way of proceeding. With no variation in prices and all prices normalized at unity we get:

$$u = v(y / d)$$

or expressed in terms of expenditure shares:

$$s_i / d = s^i(y / d)$$

where s_i is the expenditure share of good i. It is apparent that under the assumption of common scales any household buying the same expenditure share as another is necessarily at the same level of welfare. The evaluation of welfare change is in terms of the Compensating Surplus (CS). The CS is defined as the minimum change in expenditure necessary to leave the individual as well-off, prior to the change in the level of environmental amenities, as after.

DATA

Expenditure data are taken from the quinquennial Indian consumer expenditure survey for the period July 1986–June 1987 (Indian Ministry of Planning, 1991). The expenditure data refer to monthly per capita expenditure in rupees and are grouped into seven broad expenditure categories as follows:

1. Food and drink
2. Pan, tobacco and intoxicants
3. Fuel and light
4. Clothing
5. Footwear
6. Miscellaneous goods and services
7. Durable goods

Expenditure levels are reported for both the urban and rural dwellers of each state/union territory and these are further divided into different expenditure classes. The geographical coverage of the survey was the entire Indian Union excluding the Ladakh and Kargil districts of Jammu and Kashmir, rural areas of Nagaland and urban areas of Dadra and Nagar Haveli. In total 31 different states or union territories are identified (see Table 8.1). Treating each cell as an independent observation, 682 observations are available in total. The entire survey period of one year was divided into four sub-rounds each of three months duration. The samples are equally distributed throughout the year.

Mean temperature data reported in terms of monthly averages are taken mainly from Leemans and Cramer (1991). This database merges records drawn from a variety of published sources and, after various checks for quality and reliability, a terrestrial grid is created at the 0.5° level of resolution. Temperature data recorded in the original sources often incorporated adiabatic lapse rates. The presence of such factors implies that temperature values were given as if the station were located at sea level. Fortunately Leemans and Cramer report temperatures corrected to reflect mean altitude above sea level. Since the objective of the study is to determine how the cost of living changes

Table 8.1 *The States and Union Territories of India*

State or territory	Population (millions)	Major city
1 Andaman and Nicobar Is.	0.2	Port Blair
2 Andhra Pradesh	58.9	Hyderabad
3 Arunachal Pradesh	0.8	Itanagar
4 Assam	16.6	Dispur
5 Bihar	72.3	Patna
6 Chandigarh	0.4	Chandigarh
7 Dadra and Nagar Haveli	0.1	Silvassa
8 Delhi	6.2	New Delhi
9 Goa Daman and Diu	1.2	Panaji
10 Gujarat	32.2	Gandhinagar
11 Haryana	14.1	Chandigarh
12 Himachal Pradesh	4.5	Shimla
13 Jammu and Kashmir	6.8	Srinagar/Jammu
14 Karnataka	36.6	Bangalora
15 Kerala	23.7	Trivandrum
16 Lakshadweep	0.0	Kavaratti
17 Madhya Pradesh	57.2	Bhopal
18 Maharashra	62.8	Bombay
19 Manipur	1.3	Imphal
20 Meghalaya	1.4	Shillong
21 Mizoram	0.5	Aizawl
22 Nagaland	0.0	Kohima
23 Orissa	28.6	Buhubaneswar
24 Pondicherry	0.7	Pondicherry
25 Punjab	18.7	Chandigarh
26 Rajasthan	36.8	Jaipur
27 Sikkim	0.2	Gangtok
28 Tamil Nadu	50.3	Madras
29 Tripura	2.0	Argatala
30 Uttar Pradesh	119.2	Lucknow
31 West Bengal	55.8	Calcutta

Source Indian Ministry of Planning (1991)

Estimates exclude the rural population of Nagaland and the urban population of Dadra and Nagar Haveli. In total 99.8 per cent of the population of India (710.1 million people) is covered.

Table 8.2 *India: Annual Temperatures*

	State or territory	Capital	Average temperature (°C)	Range (°C)
1	Andaman and Nicobar Is.	Port Blair	26.2	2.2
2	Andhra Pradesh	Hyderabad	27.5	10.6
3	Arunachal Pradesh	Itanagar	20.3	12.5
4	Assam	Dispur	24.3	11.2
5	Bihar	Patna	26.3	16.3
6	Chandigarh	Chandigarh	24.4	20.9
7	Dadra and Nagar Haveli	Silvassa	25.8	8.0
8	Delhi	New Delhi	26.2	20.5
9	Goa Daman and Diu	Panaji	26.9	4.2
10	Gujarat	Gandhinagar	28.0	13.9
11	Haryana	Chandigarh	24.4	20.9
12	Himachal Pradesh	Shimla	15.5	15.5
13	Jammu and Kashmir	Srinagar/Jammu	16.4	23.1
14	Karnataka	Bangalora	24.9	6.2
15	Kerala	Trivandrum	26.6	2.8
16	Lakshadweep	Kavaratti	27.8	3.3
17	Madhya Pradesh	Bhopal	26.1	15.9
18	Maharashra	Bombay	26.9	5.8
19	Manipur	Imphal	19.7	10.0
20	Meghalaya	Shillong	20.1	10.4
21	Mizoram	Aizawl	20.7	8.9
22	Nagaland	Kohima	17.8	11.2
23	Orissa	Buhubaneswar	27.1	9.4
24	Pondicherry	Pondicherry	28.8	6.3
25	Punjab	Chandigarh	24.4	20.9
26	Rajasthan	Jaipur	26.4	18.4
27	Sikkim	Gangtok	12.4	12.3
28	Tamil Nadu	Madras	28.8	7.9
29	Tripura	Argartala	25.3	9.4
30	Uttar Pradesh	Lucknow	26.4	19.5
31	West Bengal	Calcutta	26.9	11.2

Note A simple average of Srinagar and Kashmir is taken for the state of Jammu and Kashmir
The record for Kavaratti is taken from the Island of Amini Divi.

with respect to climate, it is appropriate to base the analysis on averages rather than the weather during the period of the expenditure survey.

Each state/union territory was allocated a set of climate variables using the grid cell in the data set closest to the provincial capital and is therefore likely to involve some approximations, particularly where regions are large and do not possess homogeneous climates. For those states whose capital lies in a grid square not included in the Leemans and Cramer data set (Bombay, Trivandrum, Port Blair and Kavaratti) records are taken from the UK Meteorological Office (1966): for Kavaratti, however, the capital of Lakshadweep, the records taken refer to the nearby island of Amini Divi. The average temperature and temperature range of the 31 separately identified regions is summarised in Table 8.2, which displays the large differences which exist between the different regions.

The existing literature concerning the incorporation of climate variables into analyses of socio-economic behaviour such as migration decisions agrees that representing the climate by an annual average is analytically unsound, particularly where a country presents a continental type of climate with extremes of temperature. Individuals are not indifferent to the choice of locations with extremes of climate and those which offer a mild climate throughout the year.

Most interest has focused on the use of January/July averages and the use of synthetic indices such as heating and cooling degree days and the number of very hot and very cold days. The only attempt to distinguish between these different indices of climate is found in Cushing (1987) and occurs in the context of equations describing migration rates between different regions in the US. Cushing finds that the best description of climate is afforded by using January and July averages closely followed by heating and cooling degree days calculated against a 65°F benchmark. Average annual temperature performs poorly. The Indian context is somewhat different to that of the US, however, because maximum and minimum temperatures are not always in January and July and differ between regions, due in part to southern India being in the tropics.

EMPIRICAL IMPLEMENTATION

Prior to any analysis it is necessary to lend functional form to the demand equations outlined above. A highly flexible demand function for use in situations in which no variation in prices is observed is suggested by Muellbauer (1980) in his analysis of cost of living indices for demographic characteristics:

$$\frac{s_i}{d} = \alpha_i + \beta_i \frac{y}{d} + \gamma_i \frac{y}{d} \log \frac{y}{d}$$

with the adding-up restrictions that:

$$\sum \alpha_i = 1$$
$$\sum \beta_i = 0$$
$$\sum \gamma_i = 0$$

and where α, β and γ are parameters. The model is estimated using maximum likelihood techniques which take account of the non-zero correlations between the disturbances of the commodity share equations. As is customary, one equation is dropped to avoid singularity of the variance-covariance matrix and the technique is such that the results are invariant with respect to which commodity share equation is dropped.

The first method for incorporating climate variables into the scaling function d was to include a term representing the absolute deviations from some optimal temperature cumulated over 12 months. This model is extended by a term representing quadratic absolute deviations from some climatic optimum. The estimated values of the parameters contained in the scaling functions are shown in Table 8.3, where it is apparent that the coefficient on absolute deviations can be dropped without significant reduction in fit and only quadratic deviations matter. Adding further higher order terms does not result in a statistically significant improvement in fit. In this model the parameter from which all monthly deviations are calculated can be interpreted as the 'optimal climate' and takes the value of 20.3°C (69°F). That this figure should be so close to the 65°F used as a basis for calculating heating and cooling degree days is interesting, since that figure was itself derived from physiological studies rather than studies of human behaviour. The significance of the urban dummy indicates that the cost of living is 5.2 per cent higher in urban areas.

The relative improvements in explanatory power from adding the remaining three environmental variables for each commodity group are shown in Table 8.4. The improvements in fit obtained are modest. Furthermore, because the same parameters are being used to fit a system of share equations, it is not necessary that the fit of each commodity share equation should improve. The most significant improvements in fit occur for the share equations describing purchases of clothing and miscellaneous goods.

Following Cushing (1987) two alternative ways of incorporating climate variables were also considered. The first method was to include annually averaged mean temperature as both a linear and a quadratic variable together with a dummy variable for urban areas. The second method was to include the temperature of the hottest and coldest months together with the dummy variable for urban areas. These are compared with the quadratic deviations model in Table 8.5 which indicates that the method of including annual averages performs worst of all, closely followed by the method of using extreme monthly temperatures. Interestingly, this ranking differs from the one found by Cushing in that extreme monthly temperatures are not the best representation of climate. Possibly this is because even during the winter months the climate of India is typically quite warm and close to some optimum.

SIMULATING THE IMPACT OF CLIMATE CHANGE

The case for the HPF approach is clearly made: consumption patterns are affected to a significant extent by climate variables. Furthermore, not all

Table 8.3 *Scaling Function Parameter Estimates*

Variable	Parameter	Parameter
Optimal temperature	20.2250	20.3361
	(61.9721)	(78.6403)
Cumulated deviations	−0.560530E-03	
	(0.789402)	
Cumulated quadratic deviations	0.166606E-03	0.134364E-03
	(3.78961)	(9.58885)
Urban dummy	0.050597	0.051907
	(6.86753)	(7.02559)

Table 8.4 *The Relative Explanatory Power of the Various Models of Consumer Demand*

| | R^2 statistic | |
Commodity Group	Model 1	Model 2
Food	0.712	0.719
Pan		
Fuel	0.394	0.398
Clothing	0.424	0.417
Footwear	0.122	0.134
Miscellaneous	0.500	0.445
Durable goods	0.384	0.379

Table 8.5 *The Relative Performance of Different Climate Variables as Scaling Factors*

Model	Number of parameters	Log likelihood
Quadratic deviations	3	8670.56
Annual averages	3	8636.60
Extremes	3	8648.38

methods of including climate variables are equally successful. However, if one wishes to draw inferences about the amenity value of climate, the additional assumptions of demand-dependency referred to earlier are necessary. In this section the assumption of demand-dependency is made and the quadratic deviations model is used to calculate changes in the cost of living in India associated with different climate change scenarios. Rather than calculate climate change situations for each region separately, a representative climate is developed for the whole of India using the population of different regions as weights. These population weighted averages are presented in Table 8.6 and indicate that for the majority of the people for the majority of the time temperatures already exceed the optimum value of 20.3°C.

The impact of various climate change predictions for India is inferred from the general circulation model of the UK Meteorological Office. Temperature

Table 8.6 *Population Weighted Monthly Mean Temperatures*

Month	Temperature (°C)
January	19.5
February	21.9
March	26.4
April	30.3
May	32.5
June	31.2
July	28.5
August	28.0
September	28.0
October	26.8
November	23.1
December	20.1

Table 8.7 *Climate Change Assumptions for India*

| Month | Globally averaged mean temperature change | | |
	+1°C	+2°C	+3°C
January	+1.1	+2.3	+3.5
February	+1.1	+2.2	+3.4
March	+1.1	+2.2	+3.3
April	+0.9	+1.8	+2.7
May	+0.9	+1.8	+2.6
June	+0.9	+1.7	+2.6
July	+0.7	+1.5	+2.3
August	+0.7	+1.3	+2.0
September	+0.8	+1.5	+2.3
October	+0.9	+2.0	+3.0
November	+1.0	+2.0	+2.9
December	+1.1	+2.1	+3.2

Source UK Meteorological Office

increases for India are less than those expected for the world as a whole due to India's position in the lower latitudes. Rather than attempt to link these predictions to particular emission situations, it is more convenient to link them to globally averaged temperature increases. Thus, for example, when globally averaged mean temperatures increase by from 1°C to 3°C the consequences for India are described in Table 8.7.

The dates by which such changes might occur are quite uncertain, depending as they do upon future levels of greenhouse gas emissions as well as on the sensitivity of the climate to CO_2 doubling. To take the most pessimistic situation considered by the IPCC (Schimel et al, 1996), if CO_2 concentrations are not stabilized until they reach 750 ppm then the date at which globally averaged temperatures reach particular values depends entirely upon the sensitivity of the climate to increased radiative forcing. With climate sensitivity towards the lower end of that thought possible, a rise of 1°C might be attained by

Table 8.8 *Impact of Various Climate Change Assumptions on the Cost of Living in India*

Global temperature change	CS as a fraction of expenditure
+1°C	0.017433
	(10.3195)
+2°C	0.035815
	(10.4228)
+3°C	0.057903
	(10.4728)

2096. With climate sensitivity at the higher end of the range, these changes might occur by 2021, although the expected date is 2038. An increase of 2°C might be avoided altogether with a low climate sensitivity or be achieved by 2049. The expected date, however, is 2090. Finally, 3°C would be reached by 2074, but only if climate sensitivity turns out to be high.

It is noticeable that the UK Meteorological Office model, like other GCMs, predicts that temperature increases will be higher in the winter months than in the summer months. This is an important point: were the temperature increases to be uniform across all months, the impact of climate change would be all the more severe given that the empirical investigation points to quadratic rather than linear deviations from the optimal climate being more important. The pattern of climate change across the seasons matters.

Using these climate change predictions together with population weighted monthly temperatures given in Table 8.6, it is possible to predict the impact of climate change on the cost of living (see Table 8.8). If there is an increase of 1°C in globally averaged mean temperature the cost of living will increase by 1.7 per cent in India. If the temperature increase reaches 2°C then the cost of living will increase by 3.6 per cent and if temperature increases by 3°C then the cost of living will increase by 5.8 per cent. Quantitatively similar results emerge from Maddison (1997). He predicts an increase in the cost of living of 2.2 per cent for India for what amounts to a 1°C increase in globally averaged mean temperature. It is important to bear in mind that these increases in the cost of living do not account for changes in the price of marketed commodities caused by climate change. Such price changes might well occur in the case of food given the impact of climate change on agricultural productivity.

CONCLUSIONS

This study has used the HPF technique as a means of extracting information on the amenity value of climate variables from data on per capita expenditures in instances in which no price variation is observed over the data set. The approach has been applied to data describing household consumption patterns in each of 31 different regions of India. In so doing the study has addressed some of the deficiencies of earlier work, demonstrating that results cannot be ascribed to unsuitable functional forms being fitted to the data. It has also demonstrated that quadratic deviations from optimal temperatures contribute most to differences in consumption patterns.

Cost of living impacts have been derived for particular climate change situations. It appears that much of India is already beyond its climatic optimum and further increases in temperature will result in a small, but not insignificant, increase in the cost of living of between 1 and 6 per cent of expenditure. The fact that most of the warming occurs during the colder months is an offset to these increases in the cost of living.

The current analysis, however, suffers from a number of shortcomings which it would be desirable to rectify in future work. The consumption data still refer to average consumption patterns within a relatively large region. It is possible that the procedure of ascribing the climate of the capital city to the whole state or territory is deficient. Were more data to become available from smaller, climatically homogeneous regions it would be worth including a larger number of climate variables such as precipitation, sunshine and windspeed in the analysis. Similarly, the fact that no price variation is observed across the data set places restrictions on the relationships between environmental variables and marketed goods.

Appendix 1[1]

THE IMPACT OF CLIMATE CHANGE ON AGRICULTURE IN BRITAIN

THE THEORETICAL BASIS FOR RESTRICTING THE FUNCTIONAL FORM OF THE HEDONIC PRICE EQUATION

Let p represent the sale price of a 'non-standard' plot of land, $p(.)$ the hedonic price function, $c(.)$ the cost of providing structural attributes, and x_1^*, x_2^* and x_3^* represent the 'standard' level of structural attributes, locational characteristics and plot size respectively. Let 't' and 'a' represent an integer greater than one. The land owner will be prepared to sell a plot of non-standard size and non-standard structural attributes provided that:

$$p - c\left(ax_1^*\right) \geq tp\left(x_1^*, x_2^*, x_3^*,\right) - tc\left(x_1^*\right)$$

The logic of this condition is obvious. All the owners know that they can sell a standard plot with standard attributes for a profit equal to $tp(x_1^*, x_2^*, x_3^*) - tc(x_1^*)$ so that the buyer of a non-standard plot must be prepared to pay the owner at least that much plus the cost of providing the non-standard structural attributes. The perfect competition in the supply of land and structural attributes means firstly that the inequality must hold exactly and also that the derivative of the hedonic price function, with respect to the structural attributes, must be equal to the marginal cost of supplying them. This implies that:

$$p = tp\left(x_1^*, x_2^*, x_3^*,\right) - tc\left(x_1^*\right) + c\left(ax_1\right)$$

and:

$$\frac{\partial p\left(x_1^*, x_2^*, x_3^*\right)}{\partial x_1^*} = \frac{\partial c\left(x_1^*\right)}{\partial x_1^*}$$

---

1 Appendix to Chapter 3

respectively. Differentiating the equation for the price of a non-standard plot with respect to the locational characteristics yields:

$$\frac{\partial p}{\partial x^*_2} = t\frac{\partial p(.)}{\partial x^*_2}$$

In other words, whatever the functional form of the hedonic price function, the price of locational attributes must be proportional to the size of the plot. Differentiating the price function with respect to the structural attributes yields:

$$\frac{\partial p}{\partial x^*_1} = t\frac{\partial p(.)}{\partial x^*_1} - t\frac{\partial c(.)}{\partial x^*_1} + \frac{\partial c(ax^*_1)}{\partial x^*_1}$$

but by virtue of the competitive supply of attributes then the derivative of the hedonic price function is a function only of the level of structural attributes:

$$\frac{\partial p}{\partial x^*_1} = \frac{\partial c(ax^*_1)}{\partial x^*_1}$$

This necessarily implies additive separability of the hedonic price function in terms of structural characteristics. Finally, differentiating the price function with respect to the size of the non-standard plot size gives:

$$\frac{\partial p}{\partial t} = p\left(x^*_1, x^*_2, x^*_3\right) - c\left(x^*_1\right)$$

which is not a function of t. Of course t was arbitrarily chosen so that these restrictions must hold for all non-standard plot sizes.

Appendix 2[1]

THE EFFECTS OF CLIMATE ON WELFARE AND WELL-BEING IN RUSSIA

THE ASSIGNMENT OF CLIMATE VARIABLES

This Appendix describes the way in which the climate variables were constructed, how the relevant climate variables were selected, how the effect of temperature changes was computed and how the two different waves of the panel were used.

The climate variables were computed using the average measurements of 104 different weather stations in the USSR for the period between 1931 and 1960 (Müller, 1983). Each of the 113 sampling points was assigned to the nearest weather station. If the distance to the nearest weather station was greater than one third of the distance to the second nearest weather station in a different direction (at least 100° angle), the climate variables were obtained by distance-weighted linear interpolation. This whole procedure transformed the 113 sampling points into 35 different climate regions.

As the temperature in January was often negative, it was increased by 50°C. In order to keep the temperature variables consistent, all temperature variables were increased by 50°. As large regions of Russia are below sea-level, HEIGHT was increased by 50 metres. The difference in temperature between January and July was named TEMPDIF. All variables were then transformed by taking logarithms. The climate variables are thus defined from the raw variables:

JANTEMP	Average maximum day temperature in January in Celsius + 50°
JULTEMP	Average maximum day temperature in July in Celsius + 50°
TEMPAV	Average maximum day temperature in a year in Celsius + 50°
TEMPDIF	JULTEMP – JANTEMP
RAINDAYS	Average number of days on which precipitation exceeded 0.1 mm

---

1   Appendix to Chapter 6

HEIGHT   Height in metres above sea level + 50m/100
PREC   Average annual precipitation in mm
JANWIND   Average wind in January in metres/second
HUMIDITY   Relative humidity (%)
SUNHOURS   Average number of sun-hours in a year
WINTERPREC   Average precipitation from 1 October to 1 April
SUMMERPREC   Average precipitation from 1 April to 1 October

As this study only distinguishes 35 different climate areas, there was a distinct danger that our climate variables were correlated with other regional phenomena, such as differences in life-styles and agricultural practices. To counter this, our analysis first included variables denoting four big areas in Russia. Only the region of southern Russia and the Volga was significant and a dummy for it was retained in all analyses. Sampling points were assigned a rural status if they represented a village, a group of villages or a group of farms.

SELECTION

After recoding, several factor analyses were run. The first two principle factors explained 75 per cent of the variance amongst the component variables, JANTEMP, JULTEMP, RAINDAYS, RURAL, HEIGHT, WIND, TEMPDIF, PREC, TEMPAV, JANWIND, HUMIDITY, and SUNHOURS.

   These two factors were then included in the μ-equation and in the ordered-probit of the Cantril-equation, while all other climate variables were omitted. Although the factors were quite significant, inclusion of the climate variables separately explained significantly more variance than the factors on their own. This also held when four factors were retained. Adding this to the fact that Kaizer's (1974) measure of sampling adequacy was very poor (0.567), factor analysis was disgarded for determining equivalence scales.

   Expecting that climate costs were the result of a complicated interplay between climate variables, several interaction terms were created. These included the chill factor: ln(JANTEMP)×ln(JANWIND), and the stickiness factor: ln(TEMPAV)×ln(HUMIDITY). The following interaction terms were also tried: ln(TEMPAV)×ln(PREC), Rural×ln(TEMPAV), Rural×ln(PREC), ln(JANTEMP)×ln(WINTERPREC), ln(JULTEMP)×ln(SUMMERPREC), ln(TEMPDIF)×ln(PREC).

   Since there was no strong *a priori* reason for selecting a particular group of variables, a best-fit method was pursued. The selection of the variables followed the principle of maximum R^2 improvement, this being the adding and deleting of variables until the best fit at each number of regressors is found and stopping when the adjusted R^2 no longer increases.

   The same procedure was followed for the Cantril-variable. The R^2 for the model chosen was almost identical to the R^2 of the model with all variables (difference <0.001). The variables selected for the continuous Cantril model were then automatically selected for the ordered-probit Cantril model in order to keep the results comparable. The results of this procedure are shown in Chapter 6, Table 6.3. To check whether different respondents interpreted the

Cantril question in the same way, we performed the analysis separately for several education, age and gender groups (with the same climate variables). No significant difference was found for the relative values of the different intercept terms, and no qualitative difference (sign and magnitude were similar though the nil-hypothesis of equality of coefficients was rejected) was found for the variables of income and climate, suggesting that different groups of respondents interpret the question in the same way.

Wage-regressions were run and as the variables used for financial welfare fitted the wage-regression well, they are shown in Chapter 6, Table 6.2.

DERIVATION OF PROBABILITIES

In Chapter 6, Tables 6.4 and 6.6, an increase in January or July temperature is translated into half that increase for the average yearly temperature. If TEMPAV is increased by one degree, it is assumed that the temperature throughout the year has risen by one degree. Precipitation increases are also assumed to be reflected with equal precipitation-increases throughout the year. Wind is assumed not to be related to temperature or precipitation.

To derive the probabilities in Chapter 6, Table 6.6, we first compute the income gain of a change in climate for each individual. Using equation (2) for equivalence scales, this gives us an expected income gain per person and a standard deviation of that estimate. By averaging the expected gain and its standard deviation over all individuals, we obtain both an estimate of the average gain and of the standard estimate of the average gain. This yields the probabilities in Table 6.6.

By applying the approximation theorem on functions of normally distributed vectors (proved for instance in Serfling (1981) as theorem 3.3.A.), we get (β' represents the coefficients of the relevant climate vector, C_{ti} represents climate vector C at time t for individual i, S represents the covariance matrix of the relevant climate variables, whereby $t = 0$ refers to the situation without climate change and $t = 1$ refers to the changed climate conditions) for the individual expected gain:

$$e^{\frac{\beta'(C_{0i} - C_{1i})}{1-\beta_1}} - N\left(e^{\frac{\beta'(C_{0i} - C_{1i})}{1-\beta_1}},\right.$$
$$\left.\frac{(C_{0i} - C_{1i})'\,\Sigma\,(C_{0i} - C_{1i})}{(1-\beta_1)^2}\,e^{\frac{2\beta'(C_{0i} - C_{1i})}{1-\beta_1}}\right)$$

which denotes the distribution of the income gain of a climate change for an individual and

$$\frac{1}{N}\sum_{i=1}^{N} e^{\frac{\beta'(C_{0i} - C_{1i})}{1-\beta_1}} - N\left(\frac{1}{N}\sum_{i=1}^{N} e^{\frac{\beta'(C_{0i} - C_{1i})}{1-\beta_1}},\right.$$

$$\frac{1}{N^2}\sum_{i=1}^{N}\frac{(C_{0i}-C_{1i})'\sum(C_{0i}-C_{1i})}{(1-\beta_1)^2}e\frac{2\beta'(C_{0i}-C_{1i})}{(1-\beta_1)^2}\Bigg)$$

which denotes the distribution of the average gain for the sample of a climate change.

PANEL DATA

The first two waves of the Russian panel were lumped together in order to perform this analysis (two observations from one person counted as two observations from two different persons). To see whether this pooling was allowed, we tested whether, apart from the intercept term, the coefficients of the variables had changed significantly from 1993 to 1994 using a Chow-statistic. The independence was accepted at the 5 per cent significance level. Thus there was no fundamental shift in the coefficients of variables between the two waves. In order to see whether the residuals were correlated, the residuals of the first wave μ-equation were regressed on the residuals of the second wave after the coefficients of both waves had been estimated separately. The explained variance was very low ($R^2=0.04$) and the correlation coefficient was about 0.12. The same correlation coefficients were found if we estimated the models under the assumptions that all coefficients were identical between the two waves. We thus concluded that analysing the waves separately would add very little information and that lumping them together would not distort the results. As a check, the analysis for Table 6.3 in Chapter 6 was also performed using only one observation per household, whereby the second wave observation was only selected if the first wave observation contained missing values. The resulting analysis (N=3011) did not alter any of the coefficients of Table 6.3 by more than 5 per cent. Further information on the measurement of income and other variables is contained in Frijters and Van Praag (1995).

Appendix 3[1]

A HEDONIC STUDY OF THE NON-MARKET IMPACTS OF GLOBAL WARMING IN THE US

Table A3.1 *Variables and Statistics*

Variable	Definition	Mean	Standard deviation
JAN TEMP	Temperature normal in January in °C	–0.2	6.8
JAN T SQ	JAN TEMP squared	45.9	57.5
APR TEMP	Temperature normal in April in °C	12.6	4.6
APR T SQ	APR TEMP squared	179.0	120.0
JUL TEMP	Temperature normal in July in °C	24.3	2.9
JUL T SQ	JUL TEMP squared	600.0	139.0
OCT TEMP	Temperature normal in October in °C	13.8	3.9
OCT T SQ	OCT TEMP squared	207.0	113.0
JAN PREC	Precipitation normal in January in mm	66.6	47.1
JAN P SQ	JAN PREC squared	6660	9990
APR PREC	Precipitation normal in April in mm	82.7	32.9
APR P SQ	APR PREC squared	7930	5420
JUL PREC	Precipitation normal in July in mm	92.3	38.8
JUL P SQ	JUL PREC squared	10000	7390
OCT PREC	Precipitation normal in October in mm	63.4	23.5
OCT P SQ	OCT PREC squared	4570	3210
JAN P VAR	Interannual range of January precipitation in mm	151.0	90.7
APR P VAR	Interannual range of April precipitation in mm	181.0	76.5
JUL P VAR	Interannual range of July precipitation in mm	212.0	76.9
OCT P VAR	Interannual range of October precipitation in mm	186.0	67.6
JAN T VAR	Interannual range of January temperature in °C	11.5	2.0
APR T VAR	Interannual range of April temperature in °C	6.3	1.2
JUL T VAR	Interannual range of July temperature in °C	4.7	1.0
OCT T VAR	Interannual range of October temperature in °C	7.2	1.5
JAN T DIURNAL	Day-night difference of January temperature in °C	11.6	2.0
APR T DIURNAL	Day-night difference of April temperature in °C	13.8	1.7
JUL T DIURNAL	Day-night difference of July temperature in °C	13.7	2.4

---

1 Appendix to Chapter 7

Table A3.1 *(Continued)*

Variable	Definition	Mean	Standard deviation
OCT T DIURNAL	Day-night difference of October temperature in °C	14.3	2.0
HIGH SCHL	Percentage of high school graduates	70.3	10.0
COLLEGE	Percentage of college graduates	14.5	7.1
ALTITUDE	Altitude in km	.317	.359
FEMALE PART	Female participation rate in labour force	45.0	2.23
URBAN	Percentage urban	0.36	0.48
SINGLE HOMES	Percentage of housing in single family homes	69.2	10.2
OLD HOMES	Percentage of housing built before 1939	19.8	12.5
WATER CO	Percentage of units with public or private company water	69.6	22.1
POP CHANGE	Percentage change in total population	9.88	20.2
RECENT MOVES	Percentage of families recently arrived	43.0	8.12
POP	Population in millions	0.12	0.33
FEM HEAD	Percentage of families with children who have a female head	16.3	5.84
FAMILIES	Thousands of households with children	3.53	0.48
RENT	Annual rent in US$	4144	1206
SERVICE WAGE	Annual average wages for service workers	13041	3243
RETAIL WAGE	Annual average wages for retail workers	9004	1073
WHOLESALE WAGE	Annual average wages for wholesale workers	17573	4284
MANUFACTURE WAGE	Annual average wages for all manufacturing workers	19818	5458
SE	South-east regional dummy	0.33	0.47
NE	North-east regional dummy	0.11	0.32
MW	Mid-west regional dummy	0.22	0.42
RM	Rocky Mountain regional dummy	0.06	0.24
FW	Far-west regional dummy	0.06	0.23

Table A3.2 *Regional Hedonic Regressions*

Variable	Rents	Service wage	Retail wage	Wholesale wage	Manufacture wage
JAN TEMP	7.17	2.15	2.35	9.69	−4.64
	(2.09)	(0.49)	(0.99)	(2.55)	(0.79)
JAN T SQ	−0.19	−0.22	−0.26	0.33	−1.21
	(1.45)	(1.28)	(2.83)	(2.26)	(5.28)
APR TEMP	−89.8	−4.59	−32.9	−18.6	2.56
	(23.17)	(0.94)	(12.42)	(4.28)	(0.38)
APR T SQ	−2.34	−1.07	−0.81	−2.08	−1.97
	(6.35)	(2.11)	(3.09)	(4.68)	(3.01)
JUL TEMP	−10.4	−3.36	1.11	13.3	−20.6
	(2.78)	(0.71)	(0.43)	(3.18)	(3.06)

Table A3.2 *(Continued)*

Variable	Rents	Service wage	Retail wage	Wholesale wage	Manufacture wage
JUL T SQ	−0.14	0.17	0.43	1.00	0.42
	(0.55)	(0.55)	(2.46)	(3.64)	(0.87)
OCT TEMP	95.7	10.46	32.1	−8.66	7.56
	(14.61)	(1.23)	(7.04)	(1.20)	(0.68)
OCT T SQ	−0.24	0.51	0.74	1.00	4.67
	(0.46)	(0.73)	(2.05)	(1.59)	(4.95)
JAN PREC	0.00	−0.13	0.08	0.12	0.61
	(0.00)	(0.57)	(0.59)	(0.59)	(1.89)
JAN P SQ	−.005	−.002	−.004	−.003	−.001
	(6.79)	(2.28)	(7.20)	(3.59)	(0.79)
APR PREC	−0.49	−0.48	0.01	0.23	−0.54
	(2.71)	(2.12)	(0.11)	(1.22)	(1.74)
APR P SQ	.013	.003	.009	−.006	−.019
	(5.17)	(0.96)	(4.77)	(1.95)	(4.117
JUL PREC	−0.72	−0.06	−0.26	0.09	0.71
	(5.81)	(0.37)	(3.04)	(0.67)	(3.33)
JUL P SQ	.004	.005	.001	−.005	−.009
	(2.28)	(2.66)	(1.34)	(2.66)	(2.99)
OCT PREC	1.78	0.96	1.42	0.36	−2.08
	(8.50)	(3.47)	(9.81)	(1.49)	(5.76)
OCT P SQ	−.000	.000	−.004	.004	.018
	(0.15)	(0.11)	(2.36)	(1.28)	(3.98)
CONSTANT	7610.	9890.	9330.	10100.	10370.
	(109.94)	(98.03)	(181.17)	(114.97)	(77.95)
HIGH SCHL	6.65	1.80	−0.48	6.54	14.7
	(12.14)	(2.24)	(1.19)	(9.83)	(16.40)
COLLEGE	9.48	12.2	5.41	10.0	3.52
	(20.38)	(21.25)	(16.68)	(19.36)	(4.51)
ALTITUDE	−1.60	0.58	0.27	−1.35	−3.78
	(3.54)	(1.01)	(0.86)	(2.54)	(4.70)
FEMALE PART	3.65	−14.4	0.44	−20.8	−41.7
	(2.36)	(6.35)	(0.39)	(10.71)	(14.71)
URBAN	62.4	106.	28.2	121.	81.9
	(9.84)	(11.06)	(6.18)	(14.80)	(8.48)
SINGLE HOMES	−2.00	−1.14	−1.33	−1.04	2.06
	(8.59)	(3.62)	(7.74)	(3.87)	(5.14)
OLD HOMES	−4.62	−1.56	−2.82	−2.62	−3.00
	(16.30)	(4.13)	(14.04)	(8.11)	(6.57)
WATER CO	1.20	1.92	−0.18	2.03	2.62
	(7.47)	(7.81)	(1.46)	(9.98)	(10.49)

Table A3.2 *(Continued)*

Variable	Rents	Service wage	Retail wage	Wholesale wage	Manufacture wage
POP CHANGE	1.74	−0.34	0.57	−0.47	−1.66
	(11.16)	(1.58)	(4.99)	(2.53)	(5.43)
RECENT MOVE	−1.91	−4.14	−4.54	−2.78	−1.05
	(3.93)	(6.54)	(13.44)	(5.09)	(1.39)
POP	6.32	26.8	−6.87	10.4	1.67
	(3.70)	(14.19)	(5.93)	(6.40)	(0.68)
FEM HEAD	−2.40	5.75	1.98	6.54	14.5
	(4.82)	(8.77)	(5.46)	(11.56)	(18.33)
FAMILIES	3.77	−1.08	0.30	0.80	−0.74
	(7.20)	(1.48)	(0.78)	(1.17)	(0.72)
SOUTHEAST	−85.7	−10.9	15.7	−1.03	−31.3
	(7.49)	(0.73)	(1.95)	(0.08)	(1.68)
NORTHEAST	−187.	61.7	35.1	52.8	20.1
	(13.78)	(3.47)	(3.72)	(3.44)	(0.87)
MIDWEST	97.0	47.2	13.2	65.1	52.4
	(10.08)	(3.68)	(1.98)	(6.25)	(3.18)
ROCKY MOUNTAIN	−123.	−68.1	−17.5	−56.6	−38.3
	(5.89)	(2.42)	(1.20)	(2.44)	(0.97)
FARWEST	−23.8	20.0	90.3	27.0	−69.2
	(0.93)	(0.60)	(5.07)	(0.95)	(1.50)
R^2	.863	.820	.787	.846	.769
N	2179	3019	3095	2754	2358

The dependent variable is the log of average annual 1987 wages or rents
The coefficients have been multiplied by 1000 for ease of presentation
T-statistics are in parenthesis

Table A3.3 *Climate Variation Hedonic Regressions*

Variable	Rents	Service wage	Retail wage	Wholesale wage	Manufacture wage
JAN TEMP	2.36	15.3	−1.40	5.15	10.9
	(0.69)	(3.55)	(0.60)	(1.43)	(1.80)
JAN T SQ	−0.89	−0.12	−0.46	−0.23	−1.54
	(5.81)	(0.67)	(4.93)	(1.45)	(6.46)
APR TEMP	−57.2	−3.61	−20.1	−6.49	0.69
	(13.35)	(0.67)	(6.92)	(1.33)	(0.10)
APR T SQ	−2.74	−1.10	−1.22	−2.11	−2.38
	(7.60)	(2.31)	(4.96)	(4.76)	(3.93)
JUL TEMP	1.17	3.81	4.33	8.68	−17.7
	(0.33)	(0.88)	(1.79)	(2.15)	(2.82)

Table A3.3 *(Continued)*

Variable	Rents	Service wage	Retail wage	Wholesale wage	Manufacture wage
JUL T SQ	−1.61	0.13	−0.10	−0.18	−0.17
	(5.93)	(0.41)	(0.59)	(0.61)	(0.38)
OCT TEMP	33.9	−16.0	14.2	−23.3	−21.9
	(4.94)	(1.79)	(2.93)	(2.98)	(1.91)
OCT T SQ	3.36	0.21	1.64	2.88	6.07
	(5.95)	(0.31)	(4.56)	(4.41)	(6.58)
JAN PREC	−0.41	−0.63	0.53	0.91	−0.17
	(1.84)	(2.84)	(4.43)	(3.54)	(0.53)
JAN P SQ	−.005	−.002	−.004	−.003	.001
	(7.13)	(1.96)	(8.00)	(3.44)	(0.71)
APR PREC	−0.02	0.03	−0.32	0.25	−1.18
	(0.07)	(0.14)	(1.75)	(0.75)	(2.53)
APR P SQ	−.003	.006	.002	−.012	−.010
	(0.97)	(1.52)	(0.90)	(3.74)	(2.15)
JUL PREC	0.23	0.37	−0.25	−0.11	1.39
	(1.97)	(2.65)	(3.55)	(0.65)	(7.50)
JUL P SQ	.002	.000	.003	−.005	−.016
	(1.03)	(0.24)	(3.20)	(3.14)	(6.48)
OCT PREC	−1.66	0.15	−0.24	−1.19	−1.98
	(6.16)	(0.40)	(1.21)	(3.74)	(4.50)
OCT P SQ	.016	.001	.006	.013	.012
	(6.40)	(0.30)	(3.65)	(4.38)	(2.66)
CONSTANT	8360.	9940.	9620.	10300.	10100.
	(108.70)	(90.91)	(169.51)	(102.04)	(69.94)
HIGH SCHL	6.33	2.11	−0.78	7.06	15.4
	(12.12)	(2.64)	(1.97)	(10.56)	(17.68)
COLLEGE	7.92	11.7	4.78	9.39	3.13
	(17.64)	(20.74)	(15.03)	(17.71)	(4.06)
ALTITUDE	−5.01	−1.99	−0.53	−3.77	−7.08
	(12.39)	(3.95)	(1.89)	(7.96)	(10.58)
FEMALE PART	1.16	−14.2	−1.90	−24.0	−40.3
	(0.79)	(6.41)	(1.73)	(12.69)	(14.57)
URBAN	64.4	102.	27.5	121.	75.1
	(10.61)	(10.49)	(6.05)	(14.83)	(7.89)
SINGLE HOMES	−2.00	−1.51	−1.00	−1.21	1.48
	(8.98)	(4.84)	(5.92)	(4.54)	(3.81)
OLD HOMES	−4.44	−0.94	−2.56	−2.40	−2.78
	(16.62)	(2.54)	(13.35)	(7.56)	(6.51)
WATER CO	1.60	2.06	−0.01	2.19	2.85
	(10.23)	(8.36)	(0.09)	(10.57)	(11.42)

Table A3.3 (Continued)

Variable	Rents	Service wage	Retail wage	Wholesale wage	Manufacture wage
POP CHANGE	2.22	–0.27	0.85	–0.31	–1.67
	(14.90)	(1.29)	(7.64)	(1.67)	(5.62)
RECENT MOVE	–1.56	–4.44	–3.63	–2.80	–1.64
	(3.42)	(7.20)	(11.27)	(5.09)	(2.23)
POP	–2.73	27.1	5.15	10.5	2.56
	(1.50)	(13.50)	(4.32)	(5.98)	(1.01)
FEM HEAD	–2.43	4.87	2.18	6.66	13.8
	(5.15)	(7.44)	(6.29)	(12.12)	(18.30)
FAMILIES	2.12	–0.87	–0.55	0.48	0.04
	(4.26)	(1.22)	(1.44)	(0.70)	(0.04)
JAN P	0.39	–	–	–0.24	–
VAR	(6.88)	–	–	(3.47)	–
APRIL P	0.20	...	0.16	0.21	0.39
VAR	(2.96)		(3.37)	(2.52)	(3.36)
JULY P	...	...	...	0.21	...
VAR				(3.46)	
OCT P	0.50	0.20	0.25	0.17	0.22
VAR	(10.18)	(3.12)	(7.27)	(3.07)	(2.95)
JAN T	–27.9	–7.51	–10.7	–18.3	–9.69
VAR	(14.39)	(3.02)	(8.28)	(7.96)	(3.03)
APRIL T	...	15.0	–10.8	...	36.5
VAR	–	(3.49)	(4.85)	–	(7.15)
JULY T	–	–	–	–14.2	–12.4
VAR	–	–	–	(3.59)	(2.29)
OCT T	–17.9	...	6.05	17.7	11.5
VAR	(5.74)		(2.84)	(4.84)	(2.30)
JAN T	17.2	...	...	...	?
DIURNAL	(6.08)	–	–	–	–
APRIL T	–	–	–	17.4	?
DIURNAL	–	–	–	–	(3.67)
JULY T	–10.7	16.1	...	...	?
DIURNAL	(3.44)	(4.36)			
OCT T	–17.7	–18.9	–4.74	–16.3	?
DIURNAL	(4.13)	(4.98)	(3.88)	(4.11)	
R²	.874	.821	.792	.848	.777
N	2179	3019	3095	2754	2358

The dependent variable is the log of average annual 1987 wages or rents
The coefficients have been multiplied by 1000 for ease of presentation
T-statistics are in parenthesis

References

Becker, G (1965) 'A Theory of the Allocation of Time' *Economic Journal*, vol 75, pp493–517

Bigano, A (1996) 'The Amenity Value of Climate in Italy', unpublished, Department of Economics, University College London

Blackorby, C and Donaldson, D (1991) 'Adult Equivalence Scales, Interpersonal Comparisons of Well-Being and Applied Welfare Economics', in Elster, J and Roemer, J (eds) *Interpersonal Comparisons and Distributive Justice*, Cambridge University Press, Cambridge

Blomquist, GC, Berger, MC and Hoehn, JP (1988) 'New Estimates of Quality of Life in Urban Areas', *American Economic Review*, vol 78, no 1, pp89–107

Bockstael, N, McConnell, K and Strand, I (1986) 'Recreation' in *Measuring the Demand for Environmental Quality* by Braden, J and Kolstadt, C (eds), Elsevier, North Holland

Bolin, B, Doos, B, Jaeger, J and Warrick, R (1986) *The Greenhouse Effect, Climatic Change And Ecosystems*, SCOPE 29, John Wiley, Chichester

Box, G and Cox, D, (1962) 'An Analysis of Transformations (with Discussion)', *Journal of the Royal Statistical Society*, Series B, pp211–243

Bradford, D and Hildebrand, G (1977) 'Observable Public Good Preferences', *Journal of Public Economics*, vol 8, pp111–131

Brenton, P (nd) *Estimates of the Demand for Energy using Cross Country Consumption Data*, Department of Economics, University of Birmingham

Brickman, P and Campbell, D (1971) 'Hedonic Relativism and Planning the Good Society', in Apley, M (ed), *Adaptation-Level Theory: a Symposium*, Academic Press, New York

Cantril, H (1965) *The Pattern of Human Concerns*, Rutgers University Press, New Brunswick, New Jersey

Cantu, V (1969–81) 'The Climate of Italy', in *World Survey of Climatology*, Landsberg, H (ed), vol 6, pp127–172, Elsevier, North Holland

Caulkins, P, Bishop, R and Bouwes, N (1986) 'The Travel Cost Model for Lake Recreation: A Comparison of Two Methods for Incorporating Site Quality and Substitution Effects', *American Journal of Agricultural Economics*, vol 68, pp291–297

Central Statistical Office (1995) *Regional Trends*, HMSO, London

Charney, A (1993) 'Migration and the Public Sector: A Survey', *Regional Studies*, vol 27, pp313–326

Cropper, M (1981) 'The Value of Urban Amenities', *Journal of Regional Science*, vol 21, pp359–374

Cropper, M and Arriaga-Salinas, A (1980) 'Inter-city Wage Differentials and the Value of Air Quality', *Journal of Urban Economics*, vol 8, pp236–254

CSO (1995) *Travel Trends*, HMSO, London

Cushing, BJ (1987) 'A Note on Specification of Climate Variables in Models of Population Migration', *Journal of Regional Science*, vol 27, pp641–649

Deaton, A and Muellbauer, J (1980) 'An Almost Ideal Demand System', *American Economic Review*, vol 70, pp312–326

Delft Hydraulics (1990) *Strategies for Adaptation to Sea Level Rise*, Report of the Coastal Management Zone Subgroup for WMO and UNEP, Geneva

Department of Employment (1995) *The New Earnings Survey*, HMSO, London

Diamond, P and Hausman, J (1994) 'Contingent Valuation: is Some Number Better than no Number?', *Journal of Economic Perspectives*, vol 8, no 4, pp45–64

Dinar, A, Mendelsohn, R, Evenson, R, Parikh, J, Sanghi, A, Kumar, K, McKinsey, J and Lonergan, S (1998) *Measuring the Impact of Climate Change on Indian Agriculture*, World Bank Technical Paper no 402, The World Bank, Washington DC

Erickson, J (1993) 'From Ecology to Economics: The Case Against CO_2 Fertilisation', *Ecological Economics*, vol 8, pp157–175

Evans, A (1990) 'The Assumption of Equilibrium in the Analysis of Migration and Interregional Differences: A Review of some Recent Research', *Journal of Regional Science*, vol 30, pp515–531

Evenson, R, and Alves, D (1998) 'Technology, Climate Change, Productivity and Land Use in Brazilian Agriculture', *Planejemento e Politicas Publicas*, vol 18, pp223–258

Farmland Market (1994) no 42, Reed International, London

Feenberg, D and Mills, S (1980) *Measuring the Benefits of Water Pollution Abatement*, The Academic Press: New York

Focas, C, Genty, P and Murphy, P (1995) *Top Towns*, Guinness, London

Freeman, AM (1993) *The Measurement of Environmental and Resource Values: Theory and Methods*, Resources For the Future, Washington DC

Frijters, P and Van Praag, B (1995) 'Estimates of Poverty Ratios and Equivalence Scales for Russia and parts of the Former USSR', Tinbergen Institute Discussion Papers no 95–149

Gilbert, C (1985) *Household Adjustment and the Measurement of Benefits from Environmental Quality Improvements*, unpublished, University of North Carolina, Chapel Hill, NC

Goldfeld, S and Quandt, R (1965) 'Some Tests for Homoscedasticity', *Journal of the American Statistical Association*, vol 60, pp539–547

Greene, W (1997) *Econometric Analysis*, Prentice-Hall, London

Greenwood, M (1985) 'Human Migration: Theory, Models and Empirical Studies', *Journal of Regional Science*, vol 25, pp521–544

Greenwood, M and Hunt, G (1986) 'Jobs versus Amenities in the Analysis of Metropolitan Migration', *Journal of Urban Economics*, vol 25, pp1–16

Gyourko, J and Tracy J (1989) 'The Importance of Local Fiscal Conditions in Analyzing Local Labour Markets', *Journal of Political Economy*, vol 97, pp1208–1231

Hagenaars, A (1986) *The Perception of Poverty*, Amsterdam, North Holland

Halvorsen, R and Pollakowski, H (1981) 'Choice of Functional Form for Hedonic Price Equations', *Journal of Urban Economics*, vol 10, pp37–49

Hanemann, W (1994) 'Valuing the Environment through Contingent Valuation', *Journal of Economic Perspectives*, vol 8, no 4, pp19–43

Harris, J and Todaro, M (1970) 'Migration, Unemployment and Development: A Two Sector Analysis', *American Economic Review*, vol 60, no 1, pp126–142

Herwaarden, F van and Kapteyn, A (1981) 'Empirical Comparisons of the Shape of Welfare Functions', *European Economic Review*, no 15, pp261–286

Hoch, I and Drake, J (1974) 'Wages, Climate and the Quality of Life', *Journal of Environmental Economics and Management*, vol 1, no 3, pp268–295

Hoehn JP, Berger, MC and Blomquist, GC (1987) 'A Hedonic Model of Wages, Rents and Amenity Values', *Journal of Regional Science*, vol 27, no 4, pp605–620

Hossell, J, Jones, P, Rehman, T, Tranter, R, Marsh, J, Parry, M and Taylor, P (1993) *Potential Effects Of Climate Change On Agricultural Land Use And Production In England And Wales And Implications For National Policy*, Environmental Change Unit, Oxford University

Houghton, J, Jenkins, G and Ephraums, J (eds) (1990) *Climate Change: The IPCC Scientific Assessment*, Cambridge University Press, Cambridge

Howe, H, Pollak, R and Wales, T (1979) 'Theory and Time Series Estimation of the Quadratic Expenditure System', *Econometrica*, vol 47(5), pp231–1247

Hunt, G (1993) 'Equilibrium and Disequilibrium in Migration Modelling', *Regional Studies*, vol 27, no 4, pp341–349

Il Sole 24 Ore del Lunedi, '*Il Check Up delle Province*', 30 December 1991, 28 December 1992, 27 December 1993, 19 December 1994 and 18 December 1995, Milan

Indian Ministry of Planning (1991) *Sarvekshana (Journal of the National Sample Survey Organization)*, Government of India, New Delhi

IPCC (1996a) *Climate Change 1995: The Science of Climate Change*, Cambridge University Press, Cambridge

IPCC (1996b) *Climate Change 1995: Economic and Social Dimensions of Climate Change*, Cambridge University Press, Cambridge

Johansson, P-O (1987) *The Economic Theory and Measurement of Environmental Benefits*, Cambridge University Press, Cambridge

Kaizer, H (1974) 'An Index of Factorial Simplicity', *Psychometrica*, vol 39, pp31–36

Kravis, I, Heston, A and Summers, R (1982) *World Product and Income: International Comparisons of Real Gross Product*, Johns Hopkins Press, Baltimore

Landsberg, H (1969) *World Survey of Climatology*, Elsevier, North Holland

Leary, N (1994) *The Amenity Value of Climate: A Review of Empirical Evidence from Migration, Wages and Rents*, Mimeo, US Environmental Protection Agency, Washington DC

Leemans, R and Cramer, W (1991) 'The IIASA Database for Mean Monthly Values of Temperature, Precipitation and Cloudiness on a Global Terrestrial Grid' *The International Institute for Applied Systems Analysis Research Report 91–18*, IIASA, Laxenburg

Lloyd, T, Rayner, A and Orme, C (1991) 'Present Value Models of Land Prices in England and Wales', *European Review of Agricultural Economics*, vol 18, pp141–166

Maddala, G, (1977) *Econometrics*, McGraw-Hill, Singapore

Maddison, D (1997) *The Amenity Value of the Global Climate*, CSERGE Working Paper, University College London and University of East Anglia

Maddison, D and Bigano, A (1996) 'The Amenity Value of the Italian Climate', Paper presented to the Environmental Economics Forum, Royal Economic Society, 29–31 August, Ambleside, Cumbria

Maddison, D and Bigano, A (1997) 'The Amenity Value of the Italian Climate', Nota di Lavoro 11.97, Fondazione Eni Enrico Mattei

Maler, K-G (1974) *Environmental Economics: A Theoretical Enquiry*, Johns Hopkins University Press, Baltimore

Maler, K-G (1977) 'A Note on the Use of Property Values in Estimating Marginal Willingness to Pay for Environmental Quality', *Journal of Environmental Economics and Management*, vol 4, no 4, pp355–369

Math-Tech Inc (1982) *Benefits Analysis of Alternative Secondary National Ambient Air Quality Standards For Sulfur Dioxide And Total Suspended Particulates vol II*, Report to the United States Environmental Protection Agency, Washington DC

McConnell, K (1992) 'On-Site Time in the Demand for Recreation', *American Journal of Agricultural Economics*, vol 74, pp918–925

Mearns, L, Katz, R, and Schneider, S (1984) 'Changes in the Probabilities of Extreme High Temperature Events with Changes in Global Mean Temperature', *Journal of Climate and Applied Meteorology*, vol 23, pp1601–1613

Mendelsohn, R and Markowski, M (1999) 'The Impact of Climate Change on Outdoor Recreation' in Mendelsohn, R and Neumann, J (eds) *The Market Impacts of Climate Change in the United States*, Cambridge University Press, Cambridge

Mendelsohn, R and Neumann, J (eds) (1999) *The Market Impacts of Climate Change in the United States*, Cambridge University Press, Cambridge

Mendelsohn, R, Nordhaus, W and Shaw, D (1993) 'The Impact Of Climate On Agriculture: A Ricardian Approach' in *Costs, Benefits and Impacts of CO$_2$ Mitigation*, Kaya, Y, Nakicenovic, N, Nordhaus, W and Toth, F (eds), IIASA, Laxenburg

Mendelsohn, R, Nordhaus, W and Shaw, D (1994) 'The Impact of Global Warming on Agriculture: a Ricardian Analysis', *American Economic Review*, vol 84, no 4, pp753–771

Mieczkowski, Z (1985) 'The Tourism Climatic Index: A Method of Evaluating World Climates for Tourism', *Canadian Geographer*, vol 29, no 3, pp220–233

Ministry of Agriculture Fisheries and Food (1988) *Agricultural Land Classification of England And Wales – Revised Guidelines And Criteria For Grading The Quality Of Agricultural Land*, HMSO, London

Miranowski, J and Hammes, B (1984) 'Implicit Prices Of Soil Characteristics For Farmland In Iowa', *American Journal Of Agricultural Economics*, vol 66, pp745–749

Morrison, W and Mendelsohn, R (1999) 'The Economic Effect of Climate Change on US Energy Expenditures' in Mendelsohn, R and Neumann, J (eds) *The Market Impacts of Climate Change in the United States*, Cambridge University Press, Cambridge

Mortimer J (1996) 'Sweaty Socks in the Sun-Dried Shires', *The Guardian*, 6 July, Manchester

Muellbauer, J (1980) 'The Estimation of the Prais Houthakker Model of Equivalence Scales', *Econometrica*, vol 48, no 1, pp153–176

Müller, M (1983) *Handbuch ausgewählter Klimatstationen der Erde*, Gerold Richter, 3rd edition

Nordhaus, W (1990) 'Greenhouse Economics', *The Economist*, 7 July

Nordhaus, W (1996) *Climate Amenities and Global Warming*, Paper presented to the Stanford Energy Modelling Forum, Snowmass, Colorado

Ordnance Survey (1992) *The Ordnance Survey Gazetteer of Great Britain*, Macmillan Press, London

Palmquist, R (1991) 'Hedonic Methods': in *Measuring The Demand For Environmental Quality*, Braden, J and Kolstadt, C (eds), North Holland, Netherlands

Palmquist, R and Danielson, L (1989) 'A Hedonic Study of the Effects of Erosion Control and Drainage on Farmland Values', *American Journal Of Agricultural Economics*, vol 71, pp55–62

Palmquist, R B (1991) 'Hedonic Methods' in *Measuring the Demand for Environmental Quality*, Braden, J B and Kolstad, C D (editors), pp77–120, Elsevier, North Holland

Parducci, A (1995) *Happiness, Pleasure and Judgement, the Contextual Theory and its Applications*, Erlbaum Associates, Mahwah, New York

Parry, M (1990) *Climate Change And World Agriculture*, Earthscan, London

Parsons, G (1990) 'Hedonic Prices and Public Goods: An Argument for Weighting Locational Attributes in Hedonic Regressions by Lot Size', *Journal of Urban Economics*, vol 27, pp308–321

Pearce, D et al (1996) 'The Social Costs of Climate Change: Greenhouse Damage and Benefits of Control' in Bruce, J et al (eds) *Climate Change 1995: The Economic and Social Dimensions of Climate Change*, Cambridge University Press, Cambridge

Pearce, E and Smith, C (1993) *The World Weather Guide*, Helicon, Oxford

Perman, R (1994) 'The economics of the Greenhouse Effects', *Journal of Economic Surveys*, vol 8, no 2, pp99–132

Perry, A (1993) 'Weather and Climate Information for the Package Holiday Maker', *Weather*, vol 48, pp410–414

Perry, A and Ashton, S (1994) 'Recent Developments in the UK's Outbound Package Tourism Market', *Geography*, vol 79, pp313–321

Plug, E and Van Praag, B (1995a) 'Family Equivalence Scales Within a Narrow and Broad Welfare Context', *Journal of Income Distribution*, vol 4, no 2, pp171–186

Plug, E and Van Praag, B (1995b) 'Similarity in Response Behaviour Between Household Members: an Application to Income Evaluation', Tinbergen Institute, Amsterdam

Pollak, R and Wales, T (1979) 'Equity: the Individual Versus the Family Welfare Comparisons and Equivalence Scales', *American Economic Review*, vol 69, pp216–221

Pollak, R and Wales, T (1980) 'Comparison of the Quadratic Expenditure System with the Translog Demand Systems with Alternative Specifications of Demographic Effects', *Econometrica*, vol 78, pp575–612

Pollak, R and Wales, T (1981) 'Demographic Variables In Demand Analysis', *Econometrica*, vol 49, pp1553–1551

Pollak, R and Wales, T (1987) 'Pooling International Consumption Data', *Review of Economics and Statistics*, vol 69, pp90–99

Ridker, R and Henning, J (1967) 'The Determinants of Residential Property Values With Special Reference to Air Pollution', *Review of Economics and Statistics*, vol 49, no 2, pp246–257

Roback, J (1982) 'Wages, Rents and the Quality of Life', *Journal of Political Economy*, vol 90, no 6, pp1257–1278

Roback, J (1988) 'Wages, Rents and Amenities: Differences Among Workers and Regions', *Economic Enquiry*, vol 26, no 1, pp23–41

Robbins, L (1932) *An essay on the nature and significance of economic science*, Macmillan, London

Rosen, S (1974) 'Hedonic Prices and Implicit Markets: Product Differentiation in Perfect Competition', *Journal of Political Economy*, vol 82, no 1, pp34–55

Savouri, S (1989) *Regional Data*, Working Paper no 1135, London School of Economics

Schimel, D et al (1996) 'Radiative Forcing of the Climate' in *The Science of Climate Change*, Houghton, J et al (eds), Cambridge University Press, Cambridge

Seidl, C (1994) 'How Sensible is the Leyden Individual Welfare Function of Income?' *European Economic Review*, no 38, pp1633–1659

Selvanathan, S and Selvanathan, E (1993) 'A Cross Country Analysis of Consumption Patterns' *Applied Economics*, vol 25, pp1245–1259

Sen, A (1976) 'Poverty: An Ordinal Approach to Measurement', *Econometrica*, vol 44, pp219–231

Shapiro, P and Smith, T (1981) 'Preferences or Non-Market Goods Revealed Through Market Demands: In Smith, V (ed) *Advances In Applied Microeconomics*, vol 1

Sjaastad, L (1962) 'The Costs and Returns of Human Migration', *Journal of Political Economy*, vol 70 (suppl), pp80–93

Smith, J and Tirpak, D (1989) *The Potential Effects of Global Climate Change on the United States: Report to Congress*, EPA-230-05-89-050 US Environmental Protection Agency, Washington DC

Smith, K (1991) 'Recreation and Tourism' in *The Potential Effects of Climate Change in the United Kingdom*, Parry, M (ed), HMSO, London

Smith, KV (1983) 'The Role of Site and Job Characteristics in Hedonic Wage Models', *Journal of Urban Economics*, vol 13, pp296–231

Smith, V (1991) 'Household Production Function and Environmental Benefit Estimation' in Braden, J and Kolstadt, C (eds) *Measuring The Demand For Environmental Quality*, Elsevier Science Publishers, North Holland

Smith, V and Huang, J (1991) 'Can Hedonic Markets Value Air Quality? A Meta Analysis', Paper presented to the NBER Conference, Cambridge, Mass

Smith, V, Desvouges, W and Fisher, A (1986) 'A Comparison of Direct and Indirect Methods for Estimating Environmental Benefits', *American Journal of Agricultural Economics*, vol 68, pp280–290

Sohngen, B and Mendelsohn, R 'The Economic Effect of Climate Change on US Timber Markets' in Mendelsohn, R and Neumann, J (eds) (1999) *The Market Impacts of Climate Change in the United States*, Cambridge University Press, Cambridge

Steinnes, D and Fisher, W (1974) 'An Econometric Model of Intraurban Location', *Journal of Regional Science*, vol 14, pp65–80

Stigler, G and Becker, G (1977) 'De Gustibus Non Est Disputandum', *American Economic Review*, vol 67, no 2, pp76–90

Stone, J (1954) 'Linear Expenditure Systems and Demand Analysis: An Application to the Pattern of British Demand', *Economic Journal*, vol 64, pp511–527

Straszheim, M (1974) 'Hedonic Estimation of Housing Market Prices: A Further Comment', *Review of Economics and Statistics,* vol 56, no 3, pp404–406

The Times World Atlas (1992) Times Books, London

Todaro, M (1969) 'A Model of Labour Migration and Urban Unemployment in Less Developed Countries', *American Economic Review*, vol 59, pp138–148

Traill, B (1979) 'An Empirical Model of The UK Land Market and the Impact of Price Support Policy on Land Values and Rents', *European Review Of Agricultural Economics*, vol 6, pp209–232

UK Meteorological Office (1966) *Tables of Temperature, Relative Humidity and Precipitation for the World, Part V,* HMSO: London

UNDP (1995) *Human Development Report 1995*, Oxford University Press, Oxford

Van der Sar, N, Van Praag, B and Dubnoff, S (1988) 'Evaluation Questions and Income Utility', in Munier, B (ed), *Risk, Decision and Rationality*, pp77–96, Reidel Publishing Co, Dordrecht

Van der Stadt, H, Kapteyn, A and Van der Geer, SA (1985) 'The Relativity of Utility: Evidence from Panel Data', *The Review of Economics and Statistics*, vol 67, pp179–187

Van Praag, B (1971) 'The Welfare Function of Income in Belgium: An Empirical Investigation', *European Economic Review*, Spring, pp337–369

Van Praag, B (1988) 'Climate Equivalence Scales, an Application of a General Method', *European Economic Review*, no 32, pp1019–1024

Van Praag, B (1991) 'Ordinal and Cardinal Utility: An Integration of the Two Dimensions of the Welfare Concept', *Journal of Econometrics*, vol 50 (1/2), pp69–89

Van Praag, B (1994) 'Ordinal and Cardinal Utility: An Integration of the Two Dimensions of the Welfare Concept', in Blundell, R, Preston, I and Walker, I (eds) *The Measurement of Household Welfare*, Cambridge University Press, pp86–110

Van Praag, B and Flik, R (1992) 'Poverty Lines and Equivalence Scales. A Theoretical and Empirical Investigation', *Poverty measurement for economies in transition in Eastern Europe,* International Scientific Conference, Warsaw, 7–9 October, Polish Statistical Association, Central Statistical Office

Van Praag, B and Kapteyn, A (1973) 'Further Evidence on the Individual Welfare Function of Income: An Empirical Investigation in the Netherlands', *European Economic Review*, no 4, pp33–62

Van Praag, B and Kapteyn, A (1994) 'How Sensible is the Leyden Individual Welfare Function of Income? A reply', *European Economic Review*, no 38, pp1817–1825

Van Praag, B and Plug, E (1995) 'New Developments in the Measurement of Welfare and Well-Being', *The Ragnar Frisch Centennial 1995*, Oslo

Van Praag, B, Goedhart, T and Kapteyn, A (1980) 'The Poverty Line – a Pilot Survey in Europe', *The Review of Economics and Statistics*, vol 17, pp461–465

Wall, G (1992) 'Tourism Alternatives in an Era of Global Climate Change', in Smith, V and Eadington, W, *Tourism Alternatives*, University of Pennsylvania Press, Philadelphia

White, H (1980) 'A Heteroskedasticity-Consistent Covariance Matrix Estimator and a Direct Test for Heteroskedasticity', *Econometrica*, vol 48, pp817–838

Yohe, G, Neumann, J, Marshall, P and Ameden, H (1996) 'The Economic Cost of Sea Level Rise on US Coastal Properties' *Climatic Change*, vol 32, pp387–410

INDEX

OTHER EARTHSCAN TITLES

Blueprint for a Sustainable Economy
David Pearce and Edward B Barbier
'In a worthy successor to Blueprint for a Green Economy, *Pearce and Barbier provide a tour d'horizon of current thinking, and empirical results, on the challenge of developing a sustainable economy. This new book ... is an excellent "field guide" to the main issues in developing a more sustainable future'* **John Dixon, Lead Environmental Economist, The World Bank**
Bringing the arguments right up to date, this is a blueprint for the start of the century.
Paperback • £12.95 • 1 85383 515 3

Environmental Valuation
A Worldwide Compendium of Case Studies
Edited by Jennifer Rietbergen-McCracken and Hussein Abaza
The use of economic valuation methods is fundamental in the management of the environment and natural resources. This volume presents the results of a range of international applications of different valuation techniques. Among other subjects, the studies examine ways of valuing wildlife viewing, rainforest conservation, mangroves and coral reefs, rural water supplies, and controlling urban air pollution. The analysis reveals important methodological and contextual factors, highlighting key lessons and ways of strengthening future valuations.
Paperback • £19.95 • 1 85383 695 8

Economic Instruments For Environmental Management
A Worldwide Compendium of Case Studies
Edited by Jennifer Rietbergen-McCracken and Hussein Abaza
There is now key empirical evidence of the power of economic instruments in managing the environment. This volume describes the diversity of environmental problems to which a variety of economic instruments can be applied. Authoritative studies on air and water pollution, packaging, deforestation, overgrazing and wildlife management show what is needed for them to work successfully and the pitfalls to avoid in introducing them, providing valuable guidance for future applications.
Paperback • £19.95 • 1 85383 690 7

Fair Weather?
Equity Concerns in Climate Change
Edited by Ferenc L Tóth
This is a unique, cross-disciplinary assessment of fairness and equity issues in the context of global climate change – a crucial dimension in current international negotiations – written by a collection of leading scientists in economics, sociology and social psychology, ethics, international law and political science.
Paperback • £16.95 • 1853835579

Dictionary of Environmental Economics
Anil Markandya, Renat Perelet, Pamela Mason and Tim Taylor
This is the first comprehensive dictionary of environmental economics, compiled by leading academics in the field. Each expression or phrase is explained clearly in non-technical language, with references given to its use in the growing literature on the subject area. Over 1000 cross-referenced entries cover topics such as: environmental instruments for policy-making; techniques applied in environmental and natural resource economics; major issues in environmental economics and environmental management; economics of sustainable development; natural resource accounting; and international environmental agreements.
Hardback • £40.00 • 1853835293

Managing a Sea
The Ecological Economics of the Baltic
Edited by Ing-Marie Gren, Kerry Turner and Fredrik Wulff
Marine resources and fish stocks are now high on the international and economic research agendas, and the management of highly complex marine ecosystems is increasingly important. This interdisciplinary volume presents a comprehensive blueprint for managing a sea, taking into account the essential interlinked factors, such as social impacts, drainage systems, marine currents and the ecosystems involved. This is an exemplary study in the application of ecological economics to complex natural resource systems.
Paperback • £16.95 • 1853836087

www.earthscan.co.uk